International Library of Ethics, Law, and the New Medicine

Volume 60

Series editor
David N. Weisstub, Montreal, Canada

The book series International Library of Ethics, Law and the New Medicine comprises volumes with an international and interdisciplinary focus. The aim of the Series is to publish books on foundational issues in (bio) ethics, law, international health care and medicine. The 28 volumes that have already appeared in this series address aspects of aging, mental health, AIDS, preventive medicine, bioethics and many other current topics. This Series was conceived against the background of increasing globalization and interdependency of the world's cultures and governments, with mutual influencing occurring throughout the world in all fields, most surely in health care and its delivery. By means of this Series we aim to contribute and cooperate to meet the challenge of our time: how to aim human technology to good human ends, how to deal with changed values in the areas of religion, society, culture and the self-definition of human persons, and how to formulate a new way of thinking, a new ethic. We welcome book proposals representing the broad interest of the interdisciplinary and international focus of the series. We especially welcome proposals that address aspects of 'new medicine', meaning advances in research and clinical health care, with an emphasis on those interventions and alterations that force us to re-examine foundational issues.

More information about this series at http://www.springer.com/series/6224

Richard M. Zaner

A Critical Examination of Ethics in Health Care and Biomedical Research

Voices and Visions

 Springer

Richard M. Zaner
Ann Geddes Stahlman Professor Emeritus
  of Medical Ethics and Philosophy of Medicine
Department of Medicine, Primary
Vanderbilt University
Nashville, TN, USA

ISSN 1567-8008   ISSN 2351-955X   (electronic)
International Library of Ethics, Law, and the New Medicine
ISBN 978-3-319-38251-7   ISBN 978-3-319-18332-9   (eBook)
DOI 10.1007/978-3-319-18332-9

Springer Cham Heidelberg New York Dordrecht London

Printed on acid-free paper

Springer International Publishing AG Switzerland is part of Springer Science+Business Media (www.springer.com)

Preface

I should note at the outset that an early version of this book was published several years ago in Mandarin Chinese.[1] The reason for this is that a good friend, Chen-yun Tsai, Professor of Philosophy at the National Chengchi University in Taipei, asked me if I would write a book that would lay out the principal topics and themes of my several decades of work in medical humanities and clinical ethics. Since only two of my books had at that time been published in Taiwan, both including only clinical ethics narratives, he was concerned that persons in that still-developing field there should know of my philosophical views on those and other themes. His invitation came at the conclusion of a seven-lecture tour of universities in Taiwan in 2004, and it was with the many responses to and discussion of those lectures in mind that I agreed with Professor Tsai's suggestion, but only after he had agreed to undertake the book's translation into complex Chinese Mandarin, used among professionals in colleges and universities. Working with him and the editor of his university's press, Chu-po Chen, the book was in due course completed and the translation done—although, I must add, given the complexities of the language and those of translation, especially from my own not always Standard English usages, this took several years to complete.

When I decided to try and publish the book in English, however, it was clear to me that it had to be thoroughly rewritten. In the years that had passed since I undertook that project, rereading that initial text brought home to me that much of it needed rethinking and some restructuring. I am hopeful that the result, presented here, will be found to be helpful to readers.

I had already come to appreciate, as I noted in one of my early books, that the act of writing and publishing a book seems audacious. That's true, in its way; but how

---

[1] Published as *Voices and Visions: Clinical Listening, Narrative Writing*, tr. by Cheng-yun Tsai, National Chengchi University Press (NCCU), Taipei, Taiwan, 2009 (published in Mandarin Chinese only).

much more audacious, then, is writing this book, one which takes up the major parts of a lifetime of writing and attempts to put it all together in a single place? Indeed.

Still, as things have worked out, there is much that is quite new here. More than that, this is the only place where many of these ideas can be found connected together, where they are able to knock up against as well as complement each other. So, I do not hesitate at all at this time in my life to try and share these ideas with others, to give them a kind of public presence, available to one and all. I am thus happy to engage the challenge of yet another book, exploring themes that have long been a close part of my life as a philosopher making my way in the world of clinical and research medicine.

I must say, too, that several of my colleagues have played an important part in my efforts to articulate my vision and find my voice. Principal among these are Stuart G. Finder[2] and Mark J. Bliton,[3] both of them colleagues for many years at what I first set up as the Center for Clinical and Research Ethics at Vanderbilt University Medical Center. The three of us worked very closely together at both Vanderbilt and in the clinical ethics program where we were asked to set up at Nashville's St. Thomas Hospital (1991–1994). Daily conversations, joint teaching, writing, and publication went on for more than a decade, one result of which was helping me to articulate and clarify the conception of clinical ethics that finds expression here. From what I know, too, this conception has been carried over to a good many other institutions—by them, and by a number of outstanding graduate students who worked and studied with us at the Center and have subsequently become well known in their own right. Prominent among these are Tarris "Terry" Rosell, DMin, PhD, the Rosemary Flanigan Chair at the Center for Practical Bioethics[4]; Denise Dudzinski, PhD, Associate Professor and Director of Graduate Studies in the Department of Bioethics and Humanities at the University of Washington[5]; and Paul Ford, Director and Associate Professor of Bioethics, the Cleveland Clinic.[6]

I am grateful to all of them and for much of what we together achieved at when we were together.[7] Clearly, I learned much from each of them, then and later through their writings.

---

[2] Director, Center for Healthcare Ethics, Cedars-Sinai Medical Center, Los Angeles, CA.

[3] After serving as Associate Professor in the program at Vanderbilt, Mark moved to Kaiser Permanente in Los Angeles, working as head of Ethics and Applied Philosophy.

[4] Terry is also Professor at the Central Baptist Theological Seminary in Kansas City; Clinical Associate Professor, at the School of Medicine, University of Kansas Medical Center; and Clinical Associate Professor (Ethics) at the University of Kansas Medical Center.

[5] Associate Professor in Bioethics at the University of Washington, she is also Adjunct Associate Professor in the School of Law, Adjunct Associate Professor in Family Medicine, and Affiliate Faculty in the Treuman Katz Center for Pediatric Bioethics. She is Chief of the Ethics Consultation Service and Associate Chair of the Ethics Advisory Committee at the University of Washington Medical Center.

[6] He is also Director of the Neuroethics Program.

[7] There were of course many others who worked and studied with us in the Center and in various academic programs at Vanderbilt, especially as Finder, Bliton and I were active in various graduate programs.

I want also to be sure to mention my lovely wife. It has been my deepest joy to have been with her for nearly 59 years and not only for her warmth as my close companion. For during all the time I've spent in teaching and writing, she has become an impressive and important artist, working in many fields: sculpture (stone and steel), drawing and painting, soft sculpture, fashion design, and more recently becoming known and published as an important poet and fiction writer. I have been profoundly inspired by her and her many, many works. I have been most fortunate indeed to have been with her and the others for so many years.

Nashville, TN, USA   Richard M. Zaner

Contents

Chapter 1
Introduction

I should note at the outset that an early version of this book was published several years ago in Mandarin.[1] The reason for this is that a good friend, Chen-yun Tsai, a Professor of Philosophy at the National Chengchi University in Taipei, asked me if I would write a book that would lay out the principal topics and themes of my several decades of work in medical humanities and clinical ethics. He was also concerned that at that time, only two of my books of clinical ethics narratives had been published over there, and he was concerned that persons in that still-developing field in Taiwan should know of my philosophical views on those themes. His invitation came at the conclusion of a seven-lecture tour of universities in Taiwan in 2004, and it was with that prominently in mind that I agreed with Professor Tsai's suggestion, but only after he had agreed to undertake the book's translation into complex Chinese Mandarin, used among professionals in colleges and universities. Working with him and the editor of his university's Press, Chu-po Chen, the book was in due course completed and the translation begun. But given the complexities of the language and those of translation, especially from my own not always Standard English usages, this took several years to complete.

When I decided to try and publish the book in English, however, it was clear to me that it had to be thoroughly rewritten. I am hopeful that the result, presented here, will be found to be helpful to readers.

I had already come to appreciate, as I noted in one of my early books, the act of writing and publishing a book seems an audacious act. That's true, in its way; but how much more audacious, then, is writing the following book, one which takes up the major parts of a lifetime of writing and attempts to put it all together in a single place? Indeed. Yet, that is what I've done here, and what I intended to do some years ago when I was approached by Professor Tsai to write the book focused on my

---

[1] Published as *Voices and Visions: Clinical Listening, Narrative Writing*, tr. by Cheng-yun Tsai, National Chengchi University Press (NCCU), Taipei, Taiwan, 2009 (published in Mandarin Chinese only).

© Springer International Publishing Switzerland 2015
R.M. Zaner, *A Critical Examination of Ethics in Health Care and Biomedical Research*, International Library of Ethics, Law, and the New Medicine 60,
DOI 10.1007/978-3-319-18332-9_1

most developed work about medicine and health care. When I considered having it published here, one reader stated that it did little more than repeat what I'd been writing over the last decade of my active career. That, along with my conviction that much had to be done to bring the study up to more appropriate standards led me to engage in this extensive analysis.

That criticism I now regard as simply wrong-headed; there is much that is quite new here. More than that, this is the only place where these ideas can be found connected together, where they are able to knock up against as well as complement each other. So, I do not hesitate at all at this time in my life, to try and share these ideas with others, to give them a kind of public presence, available to one and all. So, I am happy to engage the challenge of yet another book, exploring themes that have long a close part of my life as a philosopher making my way in the world of clinical and research medicine.

For a number of years[2] most of my efforts were focused, in Husserl's apt terms (Husserl 1960, p. 7), on immersing myself within the world of clinical practice and biomedical research. My aim was straightforward: to understand the clinical encounter,[3] from within, as an actual participant with appertaining accountabilities and responsibilities as stringent as those that any other clinician must assume.

Before coming to Vanderbilt University Medical Center in 1981, I had already become fascinated with the complex phenomenon of clinical practice, but did not yet have the occasion to become an actual participant in clinical encounters. Like most others in this then-burgeoning field—bioethics or the medical humanities—I had been content to observe clinical situations from time to time, to think about one or another so-called 'problem' as it was presented to me, mostly by physicians, or which otherwise caught my interest, sharing my reflections mainly with colleagues in philosophy and others of the humanities and social sciences. The idea that a philosopher might actually have a legitimate place in clinical encounters, discussions and decisions had not occurred to me in any serious way until conversations with several physician friends posed the challenge—and colleagues at Vanderbilt presented the opportunity.

What it's really like to be involved as an actual participant in these therapeutic settings—especially to be and to be held *accountable* for whatever is said and done—only gradually became clear (Zaner 1994, 1995). So impressive were these encounters that I eventually decided to share what I could of the experience in several collections of narratives stemming from my clinical experiences (Zaner

---

[2] After an initial period setting up the medical humanities and social sciences program, at the invitation of Edmund D. Pellegrino, M.D., at The State University of New York-Stony Brook's medical center (1970–1973), I then set up the clinical ethics program at Vanderbilt University's Medical Center in 1981, and retired from that position in 2002 as Ann Geddes Stahlman Professor Emeritus of Medical Ethics and Philosophy of Medicine.

[3] The term 'event,' which seems to me the same as my 'encounter,' was used by Edmund D. Pellegrino as the "architectonic" principle of clinical medicine. (*See* Pellegrino 1983). My preference is for "encounter," as this seems to capture the fuller sense of its being a relationship between at least two persons, doctor and patient, but most often more than that (e.g., family, friends, and still others).

1993, 2004). In the present study, however, my aim is quite different: while presenting the core of what I have learned in my more than three decades in this field, I hope as well to engage in a number of philosophical reflections within and about the world of medicine (which could well eventually lead into a philosophy *of*, and philosophical concerns *within*, clinical medicine and biomedical research).

With the exception of Scott Buchanan's seminal effort (Buchanan 1938/1991), philosophers traditionally paid surprisingly little attention to the phenomenon of medicine—neither to theoretical issues (some of which Buchanan addressed), nor to the nature of clinical practice (which he pretty much ignored), nor to issues ingredient to medical and biomedical research (which lay largely in the future). In any case, and unfortunately, his work went largely unnoticed when it was first published and has even remained oddly obscure since its re-issue in 1991, thanks to Edmund D. Pellegrino's urging. Since the late 1960s, to be sure, an increasing number of philosophers have addressed some of the issues posed by medicine (although most have been restricted to ethics). Prior to Buchanan's study, philosophy and medicine have with only rare exceptions been out of touch with one another— exceptions include Galen's seminal attempt in the second century to synthesize the insights of Plato and Aristotle with Hippocratic medicine, and Descartes' more than casual encounters with the medicine of his time (Zaner 1982, 1988/2002).

As becomes quite evident from even a modest reading in the history of medicine, Descartes' involvements with medicine were exceptional (Lindeboom 1978). Discovering his early and enduring fascination with anatomy, physiology, and clinical therapeutics, has made a lasting impression on my reading of the history of philosophy, where one finds little if any mention of his concerns.

Why this reciprocal ignorance should have occurred, I cannot rightly say. Working within clinical and research medicine has in any case convinced me that it is intolerable, and that both medicine and philosophy are much the worse for it. There are few philosophical themes or problems whose pursuit would not greatly benefit from a serious study of medicine—especially the philosophy of science, philosophical anthropology, and epistemology, not only ethics. Similarly, most clinical or research settings within medicine would clearly benefit enormously from even modest understanding of philosophical, not to mention ethical, analysis and sensitivity.

For instance, from its inception, as several have noted, (Edelstein 1967; Leder 1992) medicine's history is rich in examples of rigorously developed funds of empirical knowledge—including a variety of methods, rules, checks, tests, theories, and other recognized features of empirical scientific endeavors, careful attention to which would be revealing and rewarding. Certainly for me, medicine has provided a wonderfully complex and fertile terrain to pursue many fundamental questions about human life: self, person, body and embodiment, sexuality, interpersonal and social relations, perception and emotion, to mention but several.[4]

---

[4] I first realized this in the course of an effort to begin making sense of my first decade of involvement in the field—and discovered that I could not do that without much deeper study of such phenomena. (Zaner 1981, in which three central themes are probed: embodiment, self, and intersubjectivity).

Ethics, of course, has captured the attention of most people alert to the involvement in health care of persons concerned with values, especially with those embedded in policy questions. Thanks to a number of developments in medicine and research since the 1960s, when ethical issues began to engage more and more philosophers and theologians, ethics has entered what seems to be a veritable renaissance. Stephen Toulmin perceptively noted some years ago that it may well be that medicine, in fact, "saved the life of ethics." (Toulmin 1982). I am fully convinced that the same could be said of many other philosophical issues, were other aspects of medical thinking and practice to attract attention to the same degree as ethics, for they offer a remarkably rich tapestry of phenomena for the philosopher, in particular for those of us who work within the philosophical discipline of phenomenology.

At the same time, physicians, now and for the foreseeable future, have begun to realize the serious need for philosophical reflection within medicine led by the efforts of Dr. Pellegrino (1970, 1974, 1979, 1983) and Dr. Pellegrino and Thomasma (1981), not only regarding the recognized ethical facets of medicine but more especially the nature of medicine itself. Other physicians, especially Eric Cassell (1976, 1984, 1991, 1997), have also for some years been grappling with these more bracing issues, resulting in a growing body of literature beginning to redress this need (Kleinman 1988; White 1988; Hunter 1991; Leder 1990; Bishop and Scudder 1990). It is clear from these writings, and from the wider medical literature, that there are numerous systematic and historical issues far beyond and in a sense even more fundamental than the more well-known involvement with ethics—issues which invite, indeed require, philosophical reflection, and that have by now received much-needed attention.

These wider issues have been at the center of my concerns from the day I first started in this field—thanks to Pellegrino's very persuasive invitation that I become 'involved,' in the common terms of the times, in 1971, as the first Director of the Division of Social Sciences and Humanities in Medicine at the new medical center of the State University of New York at Stony Brook. I quickly learned, however, that writing about what I eventually came to term the clinical encounter would have to be postponed in order to acquire experience in and understanding of clinical work. No sooner would some insight occur than, typically, it would have to be quickly revised. Little did I realize, however, that the postponement would last well over 15 years after that first, very tentative beginning.

Gradually, some ideas about clinical encounters and medicine began to be evident—thanks to what I came to call the "practical distantiation" or "circumstantial understanding" that gradually seemed to me characteristic of the philosopher's clinical involvement (Zaner 1988/2002, pp. 40, 242–48, 267–82; Charon 2006)—resulting in a number of preliminary attempts over the past several decades to articulate and probe various aspects of this complex phenomenon (Zaner 1983, 1984). Some of these ideas were worked into a later and more sustained analysis. That study, I must add, was never intended to be more than an initial exploration, and left many themes merely suggested, some poorly addressed, and others even completely unexplored. Since then, I have been self-consciously working to redress its flaws, both clinical and philosophical, and especially to begin the arduous but quite essential process of phenomenological explication (Zaner 1990, 1995, 2004).

For reasons as puzzling to me as it might for others, my essays in this field (like my earlier philosophical ones) have only rarely been published in the same place. Often as not, they appeared in publications that rarely attract wide notice (for instance, in collections of essays on special topics), especially from colleagues in philosophy. There thus seemed good reason to put these many reflections together as a sustained study, even if only some of the many topics can be addressed here.

Hopefully, this study will be of some interest to all my colleagues; more importantly, I hope that the problems analyzed will provoke others as well to turn to them. There are few enterprises in our society for which careful study has become so exigent as medicine and health care more generally. My aim in publishing these explorations is that it will contribute to this pressing need.

More broadly, and without undue pretension, I intend this study to advance the vision Pellegrino first enunciated years ago—of a "new Paideia" to which philosophy and medicine must in our times be the principal contributors, just as they jointly produced that stunning and embracing vision in ancient Greece.

At the beginning of any disciplined study of the activities of clinical ethics consultation, concerns about methodology inevitably arise. This is all the more important as I became immersed as a philosopher in the sphere of clinical work. One concept underlying the practice of clinical ethics as I have conceived it is the idea that moral issues cannot be sufficiently understood in abstraction from the situations in which these issues arise in the first place, as will be probed further in Chap. 6. What is important, in practice, is to pay close attention to the actual circumstances, the ways in which what is perceived as a "problem" has come about, and how the specific circumstances are understood by those whose situation it is. Otherwise, precisely those features of the situation, which both present problems and suggest possible resolutions, could well be missed (Chap. 3).

Another component of the method has its basis in an agreement with Alasdair MacIntyre on the relation of "virtue" and "practice." In his words, "A virtue is an acquired human quality the possession and exercise of which tends to enable us to achieve those goods which are internal to practices and the lack of which effectively prevents us from achieving any such goods" (MacIntyre 1981). This suggests that, to engage in practical reasoning about moral issues in clinical situations, is in the first instance to subject *one's own* attitudes, choices, preferences, and tastes to the standards that currently, albeit only partially, define the practice. One can attain that understanding only by actually engaging in and subsequent reflection on the practice itself. Expressed a bit differently, self-reflection is ingredient to these reflections on medicine.

Although he did not himself work within the phenomenological tradition, I find MacIntyre's view wholly consistent with certain central insights of that tradition as I've conceived these (Zaner 1981, 1988/2002, 2012). In one place, he provides a clear way to see the point here: it is only "in the course of trying to achieve those standards of excellence which are appropriate to, and partially definitive of, that form of activity" that one can understand and make judgments about its internal goods or virtues (MacIntyre 1981, p. 177). These standards refer to the practices and their evaluation:

> to understand the structure of some particular mode of practical reasoning specific to some particular social order is not only to have learned some particular way of guiding and directing action. It is also to have learned how actions are and are interpreted by others inhabiting the same social order. And that is to say, the norms of practical reasoning in any particular social order are among the norms in terms of which the presentation of each of us by him or herself in action is construed by others, so that those others may know how to respond to our action and engage in transactions with us. (MacIntyre 1987, pp. 3–4)

MacIntyre's general analysis allows us to appreciate that there are 'thick' moral concepts or virtues at work in clinical medicine. Physicians have traditionally professed to be dedicated to preserving and promoting human life and health, thereby defining their discipline according to those goods, understood as inherent to the practice. The perspective which I will elaborate in this study involves, at its basis, a moral theory of virtue in the face of vulnerability and power, based principally on what Gabriel Marcel termed *disponibilitè*—i.e., openness, truthfulness, trust, courage, insight, compassion, and candor. I assume that these (and possibly several other) virtues will be operative in any description of a clinical ethics adequate to reflective self-understanding in the clinical world (Marcel 1935, 1940, 1951).

In MacIntyre's terms, attention to the "goods internal" to every practice not only serves to delineate and at least partially specify what is essential to that practice, but such goods can be understood and given faithful expression only by experiencing, actually participating in, the practice in question. These standards necessarily invoke relationships not only with other practitioners, (MacIntyre 1981, p. 178) but also, I think, with *every* situational participant—including those needing help who are outside the practice, but on whose behalf the practitioners act: patients and their significant others (see Chap. 4). Thus, understanding the themes internal to clinical encounters involves one not only with physicians and other providers (and their various frameworks and institutions of practice), but equally with patients, their families and/or circles of intimates (see Chaps. 5 and 6).

There is a distinctive form of practical reasoning at the heart of the discipline of clinical ethics, involving a set of internally operative goods associated with the ethics consultant's skills. Based on attentive listening and cautious observing, as well as in depth conversations, these include at least: (1) identifying moral aspects of the situation from within its own particular setting and circumstances; (2) gathering or 'working up' the relevant materials and data; (3) on that basis, helping patients and their loved ones to determine and articulate, give voice to, available options; (4) helping them interpret the alternatives suggested by the available options in order to help them, those directly involved, to understand vividly both the options and their respective aftermaths; (5) and in this sense assisting them in the clarification and analysis of those decisions, so that they might then reach decisions most commensurate with their basic values and continue with their lives.

A second feature is that these skills are specific instances of a kind of moral focus that is inherent to clinical practice. On the basis of one's own practical experience, the activity of focusing and thinking about the efforts and actions specific to this practice is a type of reflective vigilance or attentiveness to its inherent "intention" (Zaner 1975). In simplest terms, this is a kind of detective work, a complex activity that might, as I indicated earlier, be termed a sort of "practical distantiation" or, in the words of Josè Ortega y Gasset, *circumstantial understanding* (Ortega y Gasset 1957). In a word, it is to reflect in the specifically phenomenological sense. In this specific sense, as will become clear in Chap. 8 and in what follows immediately below, this does not concern the therapeutic, practical thrust of the work of clinical ethics, but rather those moments when one stops to think about what has been going on. In its practical discipline, clinical ethics is concerned strictly for the unique individuals and circumstances that are involved in an encounter for their own sakes. Stopping and thinking about these situations, that is, in phenomenological reflection, the concern shifts from these individuals as such to what is exemplified by them, by their circumstances, and so on. Any individual whatever can be taken in one of these two ways: *for its own sake*, or as *an example*. The first is practical and therapeutic in its design; the latter is philosophical (see Chap. 8).

In general, then, the method begins as a descriptive phenomenology that neither reduces one phenomenon to another, nor is it interested in whether one type of affair 'causes' another; the method is decidedly anti-reductionistic. It continually emphasizes the process of discovery and the injunction of not taking routine matters for granted. To understand the ongoing process of moral experience, knowledge and decision-making in this context, we must constantly guard against the predisposition that any of these *has to* be a certain way—in particular the way a situation is initially presented.

There is nothing mysterious about this activity; it is something each of us does many times in our daily lives, with more or less skill and attention. The method, Edmund Husserl at one point observed, is the very same as that which "a cautiously shrewd person follows in practical life wherever it is seriously important for him to 'find out how matters actually are'." The point is that, regarding whatever may be at issue, the "seriously shrewd" person must judge on the basis of sound evidence—in Husserl's words, "on the basis of a giving of something itself, while continually asking what can be actually 'seen' and given faithful expression" (Husserl 1929/1969, pp. 278–79). Where an inherent intention is clinical and practical—whether it be a matter of a physician attempting to diagnose and manage an infant's illness, or a family trying to understand what's going on with their baby—to seek sound judgment requires a concerted effort to know, really and truly, just how matters actually stand. At the outset, clearly, nothing must be taken for granted; hence the serious need for cautious listening, observing, and focused conversations (Zaner 2012).

The work of clinical ethics is focused on these or those individuals for their own sakes, and is directed primarily toward empowering those whose situation it is, helping them to reach their own decisions with the fullest understanding possible within the limits, constraints, and resources available within their actual circumstances (Zaner 1981, pp. 175–216). The ethicist's work is therefore designed in its own way as *therapeutic*: through careful observing and listening, helping participants identify what is at issue for each person; helping each become reflectively aware of and think about their respective moral frameworks; delineating, weighing, and imaginatively probing the available alternatives most consonant with those moral frameworks; and helping each attain clarity about the "stakes" and decisions so as to enable them to live with the aftermaths.

To enter into any ongoing clinical situation is inevitably to find oneself as also a participant, hence to be enmeshed in the different "stakes" ingredient to and defining the situation. To conceive this role as empowerment, is to recognize that the ethics consultant must be quite as *accountable* (and *held* accountable) as any physician or other clinician. To be sure, this accountability must be *appropriate*: *for* what is and is not said and done, and *to* those whose situation it is most immediately. As I will repeatedly emphasize in this study (see Chaps. 4 and 7), the idea of responsibility is thus central to clinical ethics: to be *responsible for* what is and is not said and done, and *responsive to* those persons whose situation it is.

The practice of a clinical ethics should be seen as a *process* of helping people come to an understanding of their own moral beliefs and in the course of that, to find out what decisions they can live with and which decisions they can act upon. This process requires that practitioners enable the other participants to find out, first, what their deepest values are and secondly, whether or not these values conflict with elements of the aftermath the participants feel they can live with.

In this process it is certainly plausible that the emotions experienced, and the meanings which accompany those experiences, evolve into *particular* meanings that can provide evidence of one's deepest values. If it is true to say that the *specific* characteristics of these experiences are context-dependent, then it is entirely plausible to say that these meanings will be subject to the limitations imposed by the variability of that context. The result is that morally relevant terms in all cases need to be understood in relation to the discovery of the particular *meanings* of the participants' experiences within the situation, that is, as they are understood by those persons whose situations it is.

Providing this kind of help requires disciplined self-knowledge: frequently practiced and disciplined reflection intended to delineate one's own feelings and commitments, moral beliefs and social framework, followed by a rigorously disciplined suspension of those features, in order to understand what things are like for the other participants—a kind of practical distantiation that undergirds the activity of compassion or affiliative feeling.

Specifically with regard for the injunction not to take routine matters for granted, in clinical situations it is frequently necessary to re-evaluate moral attitudes in order to clarify, as much as possible, a person's ability to respond to the contingent and indeterminate character which pervades the most difficult medical cases. The point is one suggested much earlier by John Dewey:

> What is needed is intelligent examination of the consequences that are actually affected by inherited institutions and customs, in order that there may be intelligent consideration of the

ways in which they are to be intentionally modified in behalf of a generation of different consequences. (Dewey 1981, pp. 217–18)

Accordingly, the reasoning employed should be pragmatic, by which I mean that it should have a method that takes into account the range and depth of the context influencing the values expressed in concrete circumstances. It is precisely because the orientations in clinical encounters are both complex and goal-oriented that this pragmatic attitude is one modality that is expected and practiced by those involved with medicine. In this sense, a phenomenologically informed form of pragmatism illustrates the importance of understanding relationships as purposive intentions. Such an understanding also provides a purchase on the idea that our attitudes towards the intentional features of these relationships are co-determinate for both the meaning of our current values and the orientations toward those values that we intend to experience in the future, as John McDermott has pointed out (McDermott 1986).

Every situational participant brings to the encounter his/her own autobiographical situation: (Schutz and Luckmann 1973) typifications, life-plan, undergirding moral and/or religious framework, etc. These encounters are also socially framed by cultural values, professional codes, governmental regulations, hospital policies, unit or departmental protocols, etc.—any or all of which may contribute to "what's going on" in any specific case. And, given the understanding of experiential and contextual relevance, any of these should be open to "intelligent consideration of the ways in which they are to be intentionally modified in behalf of a generation of different consequences." In the light of this understanding and its meaning, my aim here is to emphasize that the concept of the "pragmatic" refers to the idea that moral reasoning actually consists in our attempts to harmonize our perceptions of the specific situation, our moral sensibility as it manifests itself in that situation, and any excellences and goods which might help directly or indirectly to explain those perceptions and sensibilities. And, as I will underscore later, this brings every clinical ethics consultant into the embrace of responsibility and accountability (Chap. 7).

It is wholly reasonable that this method requires that one remain sensitive to the specific components of particular cases, for this is just what allows for adjustment to the moral complexities of each case without disrupting existing procedures and principles in ways that would yield little benefit. For instance, clinical diagnosis, specifically in its 'detective work' of methodically delineating alternative therapies, no less than the ongoing conversations with parents, turns out to exhibit features strikingly similar to this methodology. Thus, no matter how complex a case becomes, the relevant decision makers can always 'think through' the relevant concerns, as well as the 'possibly otherwise,' by way of some reasoned reference to both the evidence given in the immediate case under consideration and the rationales associated with outcomes of other previous cases.

The salient point of the phenomenological elements of a 'pragmatic phenomenology' is that phenomenological analysis reveals to us, from an intersubjective perspective, persons as subjects capable of projecting ends and engaging in purposeful actions to achieve those ends. This perspective is particularly important when considering the parents of sick or injured children, as well as grievously afflicted adults, because such analysis also shows that persons can simultaneously hold *relationships* with a multiplicity of concerns and purposes that are *experienced* as being equally as important, *to them*, as the epistemic qualities posited in a medical viewpoint.

What this method of circumstantial understanding, or as I've also termed it, interpretive possibilizing, (Zaner 2012) enforces is the idea that any previous medical response that demanded a moral decision can change, therefore such a decision requires a renewed, persistent and reflective monitoring in order to certify whether or not the previous rationale continues to address all the relevant elements of the situation.

In what follows, I will try to make these ideas as clear as I possibly can while addressing a specific range of ideas that have come to light in the course of my efforts to establish an ethics that is genuinely responsive to the demands and constraints of clinical encounters. To do this, brief journeys into certain aspects of the history of this general field of endeavor will prove necessary from time to time, so as to provide the particular issues their specific contexts. Most especially will it be essential to make the idea of a philosophy concretely engaged in clinical work something that is both evident and reasonable.

References

Bishop, Anne H., and John R. Scudder Jr. 1990. *The practical, moral, and personal sense of nursing: A phenomenological philosophy of practice*. Albany: State University Press of New York.

Buchanan, Scott. 1938; reissued 1991. *The doctrine of signatures: A defense of theory in medicine*, ed. Peter P. Mayock Jr., Foreword Edmund D. Pellegrino. Urbana/Chicago: University of Illinois Press.

Cassell, Eric J. 1976/1985. *The Healer's art: A new approach to the doctor-patient relationship*. Cambridge, MA: MIT Press. *Talking with patients*, two volumes, Cambridge, MA: MIT Press.

Cassell, Eric J. 1984. *The place of the humanities in medicine*. New York: The Hastings Center Publications.

Cassell, Eric J. 1991. *The nature of suffering and the goals of medicine*. New York/London: Oxford University Press.

Cassell, Eric J. 1997. *Doctoring: The nature of primary care medicine*. Oxford: Oxford University Press/Milbank Memorial Fund.

Charon, Rita. 2006. *Narrative medicine: Honoring the stories of illness*. New York: Oxford University Press.

Dewey, John. 1981. *The quest for certainty: The later works*, vol. 4, 217–218. Carbondale: Southern Illinois University Press.

Edelstein, L. 1967. *Ancient medicine*. Baltimore: The Johns Hopkins University Press.

Hunter, Kathryn Montgomery. 1991. *Doctors' stories: The narrative structure of medical knowledge*. Princeton: Princeton University Press.

Husserl, Edmund. 1929; 1969. *Formal and Transcendental Logic*. Trans. D. Cairns. The Hague: Martinus Nijhoff.

Husserl, Edmund. 1960. *Cartesian Meditations*. Trans. D. Cairns. The Hague: Martinus Nijhoff.

Kleinman, Arthur. 1988. *The illness narratives: Suffering, healing and the human condition.* New York: Basic Books, Inc.

Leder, Drew. 1990. *The absent body.* Chicago: University of Chicago Press.

Leder, Drew (ed.). 1992. *The body in medical thought and practice.* Dordrecht/Boston: Kluwer Academic Publishers.

Lindeboom, G.A. 1978. *Descartes and medicine.* Amsterdam: Rodopi NV.

MacIntyre, Alasdair. 1981. *After virtue: A study in moral theory.* Notre Dame: University of Notre Dame Press.

MacIntyre, Alasdairs. 1987. Practical rationalities as forms of social structure. *Irish Philosophical Journal* 4(1–2): 3–19.

Marcel, Gabriel. 1935. *Être et avoir.* Paris: F. Aubier.

Marcel, Gabriel. 1940. *Du refus à l'invocation.* Paris: Gallimard.

Marcel, Gabriel. 1951. *Le Mystère de l'être,* tomes 1 et 2. Paris: Éditions Montaigne.

McDermott, John J. 1986. Pragmatic sensibility: The morality of experience. In *New directions in ethics: The challenge of applied ethics,* ed. Joseph P. DeMarco and Richard M. Fox, 113–134. New York: Routledge and Kegan Paul.

Ortega y Gasset, José. 1957. *Man and people.* New York: W. W. Norton and Company.

Pellegrino, Edmund D. 1970. The most humane of the sciences, the most scientific of the humanities. In *The sanger lecture.* Richmond: Medical College of Virginia, Virginia Commonwealth University.

Pellegrino, Edmund D. 1974. Medicine and philosophy: Some notes on the flirtations of minerva and aesculapius. In *Presidential address.* Philadelphia: Society for Health and Human Values.

Pellegrino, Edmund D. 1979. *Humanism and the physician.* Knoxville: University of Tennessee Press.

Pellegrino, Edmund D. 1983. The healing relationship: The architectonics of clinical medicine. In *The clinical encounter: The moral fabric of the physician-patient relationship,* ed. E.A. Shelp, 153–172. Boston/Dordrecht: D. Reidel Publishing Company.

Pellegrino, Edmund D., and David C. Thomasma. 1981. *A philosophical basis of medical practice: Toward a philosophy and ethic of the healing professions.* London/New York: Oxford University Press.

Schutz, Alfred, and Thomas Luckmann. 1973. *Structures of the life-world.* Evanston: Northwestern University Press.

Toulmin, Stephen. 1982. How medicine saved the life of ethics. *Perspectives in Biology and Medicine* 25(4, Summer): 736–750.

White, Kerr L. (ed.). 1988. *The task of medicine: Dialogue at Wickenburg.* Menlo Park: The Henry J. Kaiser Family Foundation.

Zaner, R.M. 1975. On the sense of method in phenomenology. In *Phenomenology and philosophical understanding,* ed. E. Pivcevic, 125–141. London: Cambridge University Press.

Zaner, R.M. 1981. *The context of self: A phenomenological inquiry using medicine as a clue.* Athens: Ohio University Press.

Zaner, R.M. 1982. The other descartes and medicine. In *Phenomenology and the understanding of human destiny,* ed. S. Skousgaard, 93–119. Washington, DC: University Press of America/ Center for Advanced Research in Phenomenology, Inc.

Zaner, R.M. 1983. Flirtations or engagements? Prolegomenon to a philosophy of medicine. In *Phenomenology in a pluralistic context,* ed. W.L. McBride and C.O. Schrag, 139–154. Albany: State University of New York Press.

Zaner, R.M. 1984. Is 'Ethicist' anything to call a philosopher? *Human Studies* 7(1): 71–90.

Zaner, R.M. 1988/2002. *Ethics and the clinical encounter.* Lima: Academic Renewal Press. (Reprinted from Englewood Cliffs: Prentice-Hall, Inc.).

Zaner, R.M. 1990. Medicine and dialogue. In Special issue: "Edmund Pellegrino's philosophy of medicine: An overview and an assessment", ed. H.T. Engelhardt. *Journal of Medicine and Philosophy* 15(3, June): 303–325.

Zaner, R.M. 1993. *Troubled voices: Stories of ethics and illness.* Cleveland: The Pilgrim Press.

Zaner, R.M. 1994. Phenomenology and the clinical event. In *Phenomenology of the cultural disciplines*, Contributions to phenomenology, vol. 16, ed. M. Daniel and L.E. Embree, 39–66. Dordrecht/Boston/London: Kluwer Academic Publishers.

Zaner, R.M. 1995. Interpretation and dialogue: Medicine as a moral discipline. In *The prism of self: Essays in honor of Maurice Natanson*, ed. S. Galt Crowell, 147–168. Dordrecht/Boston: Kluwer Academic Publishers.

Zaner, R.M. 2004. *Conversations on the edge: Narratives of ethics and illness*. Washington, DC: Georgetown University Press.

Zaner, R.M. 2012. *At play in the field of possibles: An essay on free-phantasy method and the foundation of self*. Bucharest: Zeta Books.

Chapter 2
Themes and Schemes: A Prelude

I should confess at the outset that each time I have set out to write something introductory about the medical humanities, even more about clinical ethics, the attempt seems to take on a life of its own, proceeding through undergrowths of first one then another set of bristling themes, questions and issues, leaving me unable to decide what to include and what to leave out. The point is that all of these issues, questions and themes seem to me essential for gaining any serious understanding of this apparently unruly field, the medical humanities or bio-ethics, especially in its clinical involvements. The point is also that it will take some time for me to work my way into these themes and schemes that lie beneath them. I ask for patience, then, to allow me to begin this chore, which I continue to regard as among the more important of our times.

2.1 Preliminary Reflections on Themes

Writing about phenomenology and medicine some years ago, I emphasized a point that I believed to be fundamental to understanding medicine. "Medicine's central theme is clear: the clinical event governs" (Zaner 1997, p. 446).

I came to the same view in other writings, (Zaner 1994a, b, c, 2005) where I urged that not only ethical issues but more broadly phenomenological matters closely related to the clinical event are similarly fundamental to understanding medicine: the interpretation of symptoms, (Zaner 1988/2002) clinical judgment, (Zaner 1994a, b, c) the social structure of clinical encounters, (Zaner 2006) the multiple forms of responsibility and uncertainty, (Zaner 1996) and others. The focus on the clinical event suggests that the core phenomenological theme is the health care professional-patient relationship, and thus the central epistemological theme of medicine follows directly from its practical, clinical orientation: medical knowledge is ordained to the goal of helping afflicted and compromised persons;

© Springer International Publishing Switzerland 2015
R.M. Zaner, *A Critical Examination of Ethics in Health Care and Biomedical Research*, International Library of Ethics, Law, and the New Medicine 60, DOI 10.1007/978-3-319-18332-9_2

such knowledge, in this sense, has its root in *practice*. In this, we come across another dimension of the pragmatic character of my study.

A survey of the main issues in both public and professional settings since the early 1960s, suggests that almost every one of them takes its point from its relation to, or implications from, that special relationship between professional and patient: the clinical encounter. This is as true for solid organ transplantation as it is for in utero surgery, as true for prevention as it is for prognosis, as true for dying persons as it is for embryos. It is also true even for the more exotic topics occasioned by the continuous outpouring of procedures, drugs and equipment from medicine's expanding bounty, whether diagnostic or therapeutic.

Hans Jonas had persuasively argued that "practical use is no accident but is integral to [modern science]... science is technological by is very nature" (Jonas 1966). The very same characteristic seems true of medicine as well—not only in view of its acceptance of this science as the major feature of medicine, but even more, as suggested, its central commitment to providing understanding (in the form of a diagnosis) and treatment for individual persons.

It is in any event clear that the primary issues of interest to philosophers and ethicists more generally have been the practical, ethical, and epistemological facets of medicine: clinical judgment in diagnoses (what is wrong?), therapeutics (what *can* be done about it?), and decision-making (what *should* be done?). In the same way, there has been serious interest in determining the nature of clinical judgment, (Pellegrino 1979a, b) the structure of clinical encounters, (Cassell 1979/1985, 1985) and the illness experience (Frank 2001, p. 241; Kleinman 1988).

Although this focus on the clinical event brought out a number of serious moral issues, for the last five or six decades of the involvement of philosophers and others in the humanities in health care, the questions that captivated both public attention and professional study had to do mainly with what may be called issues at the end of life. Popular and professional publications alike, as well as government agencies, became absorbed with such questions as aid in dying (by physicians or others), euthanasia, brain death and, along with these, advanced directives (living wills, durable power of attorney for health care). When the Nancy Cruzan case reached the U. S. Supreme Court in 1990,[1] not only did the right to sign an advance directive become securely established, but the case motivated Congress to pass, in 1991, a law requiring all health care institutions duly to inform all incoming patients and their significant others of this right to refuse treatments when terminally ill. The only remaining questions were about public and professional education—which proved substantial indeed.

---

[1] Nancy Cruzan was a 26 year-old woman who suffered a single car accident in Missouri and, resuscitated by the emergency personnel who found her partially submerged in water, she eventually was diagnosed as persistent vegetative state. Her parents finally decided to ask courts permission to have the only life-support being used, a feeding tube, removed and allowing her to die. Opposed by so-called "right to life" groups, she did manage to die. Her parents had taken their request through the Missouri court system and then to the U. S. Supreme Court.

Other difficult, even harsh, questions had, of course, already occasioned heated disputes: abortion, treatment for severely premature babies, ensuring informed consent and privacy, allocation of scarce resources, and still others. Some became part of our public iconography: a liver transplant for the 3-year-old child living in poverty; choosing who should get renal dialysis when there are not enough machines; what to do about exotic, novel alternative forms of pregnancy (surrogate mothers, artificial insemination, in vitro fertilization and embryo transfer, stem cell research, human cloning, etc.); whether family planning should include sex pre-determination; and many others. Dramatic, highly sophisticated types of diagnostic imaging technology brought on still more issues: ultrasound (US), computer-assisted tomography (CAT), positron emission tomography (PET), magnetic resonance imaging (MRI), and so on. Others arrived in the wake of breathtaking forms of treatment: multi-drug chemotherapy, cellular transfusion, mind-altering drugs, and the like. Still others came on the heels of sensational forms of surgical intervention: intrauterine surgery, solid organ transplantation, neural cell implantation, stem cell infusion, etc. And some came about simply because the health care system came to the point of near-collapse from mal-distribution of resources; or from too many non- and under-insured people; or other sources stemming from the broader society's inability to confront and resolve grievous social problems such as the increasing numbers of children living in poverty, ghettos, damage from violent weather patterns, and wide-spread hunger. Perhaps the most compelling irony of the latter half of the twentieth century in the United States is the existence of a society whose wealth and power exceeded those of other nations yet could not properly feed, clothe, house, educate or employ many of its own citizens—and thus recently wound up far below other nations in such matters as infant mortality, care for difficult pregnancies and others.

Beyond these concerns, questions of social justice (equality, diversity, distribution, violence, disability, etc.) continue to plague most nations in the new millennium. Nevertheless, a number of other issues have become prominent and, doubtless, will complicate social and political discourse even more. Whether questions of social justice are addressed in fruitful ways or not, these other, already conspicuous matters seem especially compelling.

Unlike the preoccupying questions of the past five decades (withdrawal and withholding life supports, do-not-resuscitate orders, living wills, death and dying, right to die, aid-in-dying, and other end of life issues), those that have already begun to enthrall and even obsess so many people and institutions concern the opposite end of the life-spectrum: life before birth—from the impact of molecular biology and genetics on medical theory and clinical practice, to questions of the legitimate use of genetic information, cellular manipulations, embryo research, to pre-natal diagnosis and fetal interventions more broadly.[2]

---

[2] For instance, I became deeply involved in the very first fetal surgical intervention protocol at Vanderbilt University Medical Center, when I was asked to develop an informed consent and consultation procedure for it. It was designed for in utero closure of fetal lesions due to spina bifida when diagnosed as early as 20 weeks gestation. This was initiated in 1997 and continued until my retirement in 2001, subsequently becoming a full-fledged NIH-sponsored protocol at several institutions.

The implications especially of the genome project have clearly stretched our moral imagination well beyond traditional limits. If a central, governing thought in the late twentieth century was the "what-is-to-come" (the future: *Zukunft*)—that is, death, believed to be determinative for the being and life of humans—the twenty-first century is already noticeably turning the other direction in search of religious, moral and ontological guidance. Birth, not death, seems definitive; genetics, not geriatrics, is seen as the basic medical and biomedical discipline; not end-of-life, but life-before-birth has become focal. With that critical shift, questions invariably emerge about the relationships between medicine's practices (clinical encounters) and its pervasive commitment to being a science, or dependent on the biomedical sciences. Indeed, with the latter's full emergence since the end of World War II, the sense of 'clinical' has begun to fade, although by no means has it disappeared; better expressed, 'clinical' has increasingly become transformed, now being more and more regarded as a matter of 'science,' as in the emergence of so-called 'evidence-based' medicine (See Henry et al. 2007).

If we would thus seek to explicate the phenomenon of medicine phenomeno-logically, especially in its more recent transformation through innovative develop-ments in biomedical science, we have no choice but to attempt precisely what Edmund Husserl demanded in the first pages of his great work, *Cartesian Medita-tions*. There, he presented the required project bluntly: asking how the idea of science with its goals and methods can be "uncovered and apprehended," he answered unequivocally that "there is nothing to keep us from 'immersing our-selves' in the scientific striving and doing that pertain to them." That is, we can and must ask ourselves in the most rigorous manner what it is that the 'scientist' is truly attempting to do in that intellectual labor called 'science;' what are scientists after and how will they know when they have arrived? We must do this rigorous kind of reflective thinking, Husserl insists, "in order to see clearly and distinctly what is really being aimed at" (Husserl 1960, p. 9). Precisely this act must be attempted if we would similarly disclose the inner sense of medicine and its methods. And this kind of work is essential, I believe, to the work of clinical ethics and, more broadly, the medical humanities.

Following that effort is what leads to the disclosure of the phenomena noted above: the sense of 'symptoms' and the variety of possible 'interpretations,' the explication of diagnosis and clinical judgment as well as the illness experience, the modes of givenness specific to the moral dimensions of clinical encounters, and the place of uncertainty and error, as well as the phenomena of trust and responsibility (see Chap. 4), and still others. Explicating these, furthermore, requires careful attention to their temporality structures, as well as the subjective (noetic) and objective (noematic) aspects, the ego dimensions and belief character appertaining to the way in which these are experienced in times of illness and of healing. For instance, not only is illness concretely experienced as a disruption in the flow of daily life, but at the same time it discloses an often unnoticed dimension of gratitude and the promise of being healed and returning to normal life—that helps make prominent, too, the multiple forms of inner time awareness specific to illness: the struggle with 'feeling bad' through the temporal modes of 'waiting to heal' or for scars to form.

Now that fundamental changes are occurring at the very heart of the new union of medicine with biomedical science, however, it is imperative to undertake with renewed energy those very efforts again: to immerse ourselves in the newly emerging currents of medicine that are only now becoming clear enough to permit that immersion and explication. That such phenomenological explication from within medicine itself must take nothing for granted goes, hopefully, without saying. That the course of the analysis must demand clear evidence for each judgment about each new prominence as it appears is also obviously true. What is occurring, it must be emphasized, is so fundamental—and is happening with such amazing speed within medicine, arguably among the most significant of human endeavors—that extraordinary caution and discipline must take precedence. As Hans Jonas once urged in a related context, "Since no less than the very nature and image of man are at issue, prudence becomes itself our first ethical duty, and hypothetical reasoning our first responsibility" (Jonas 1984, p. 141). This statement I would qualify only modestly, emphasizing not so much "prudence," which stems ultimately from Aristotle and has its place in this work, but accentuating instead another Greek notion key to the history of medicine and philosophy: self-restraint (*sophrōsyne*) as fundamental to clinical and, as will hopefully become clear as we proceed, phenomenological responsibility.

2.2 Issues Remain Complicated, Often Opaque

Matters are made all the more complicated by the fact, as I see it, that not very much about the discipline of ethics in medicine is well understood—by the general public, medical personnel, well-meaning academic colleagues, and even by some of the medical humanities' more dedicated practitioners. This is even truer when the new developments in the world of medicine and biomedical research are taken into consideration—as they must be, in order to make sense of the ethical issues these novel developments bring about. Indeed, the more serious are the latter, the more are they the occasion for many of the persistent questions, the intractable conflicts, and unending disputes integral to the new genetics and biomedicine.

Put this over against the amazing growth and popularity of the field of bioethics—from meager beginnings in the early 1960s to the world-wide status it enjoyed only a decade later—and we face one of the most intriguing questions about this field: What can account for this striking popularity of an endeavor still so poorly explained, especially in what may seem such central and important questions about human life and death? Equally perplexing, it remains unclear just what the at times raucous concern is all about (Zaner 2003, 2013).[3] And that, to be candid, is

---

[3] I think especially the continuing disputes over abortion, persistent vegetative state and, more recently, stem cell research.

complicated by the further fact, as I'll point out later on, that few of its practitioners and advocates agree on what 'it' really is! Bioethics, like ethics more broadly, remains seriously problematic (MacIntyre 1981; Toulmin 1982).

Even after I had been involved for more than a decade, it was very common to face the question—it seemed at the time more accusation than query—"what the hell is that philosopher doing in our hospital?" (Zaner 1994a, b, c). Or, as I have often had to wonder: "Is 'ethicist' anything to call a philosopher?"—a question that seemed imperative to state openly. This, too, appeared to be more an ill-mannered indictment from colleagues in the humanities than a question seeking serious response.

I must also acknowledge that I am fully cognizant that there are some who apparently have a firmer grip than I on the what this field is all about; nor do they seem nearly as puzzled as I. Robert Veatch, for one, published a book a long time ago with what still seems to me an audacious title, *A Theory of Medical Ethics*, (Veatch 1981) an imprudent undertaking, I have thought, simply because, as subsequent events have made plain, it was far too early for anyone to have the necessary understanding of everything about this field, even whether this was in any sense a 'field'. Tom Beauchamp and James Childress put out another, very popular, if also, as I think, premature book, *Principles of Medical Ethics* (Beauchamp and Childress 1983/2012). A bit more modestly, H. Tristram Englehardt, Jr. waited for a few years before publishing his own version of such grand and unreserved visions; (Englehardt 1986) and there are others, (Katz 1984; Pellegrino and Thomasma 1987) no one of which could be 'the correct' version, as each most often disagreed more or less seriously with the others—a phenomenon not in the least uncommon in our times, as another commentator noted early on as well.

Nor should I forget to mention the vast number of conferences, professional meetings, symposia and the like, along with their resulting published collections of essays (to which, I confess, I have also contributed now and then), not to mention the many other anthologies and book series, which appeared over this span of time, and continue to appear. Bioethics, or the medical humanities, has been a prolific growth industry, not only in the United States, but worldwide. Not to press the point too much, most of this vast literature exhibits a marvelous confidence about the field, its so-called main problems, methods, and even what theories or principles are thought to be appropriate for dealing with those issues.

I confess to having been as perplexed by this striking production as I have been at its popularity. What's more, I still remain somewhat at a loss; at times I am convinced that I have neither the wits nor the patience to appreciate, or even keep up with, all these books, essays, articles, speeches, conferences, presentations, and the rest, despite the fact that I have myself been seriously involved since 1971, have participated in quite a few of these goings-on, and before and after my retirement in 2002 taught undergraduate, graduate, medical, nursing, law, religious studies, and other courses and seminars for many years—all of them at least related to this arena of themes and questions. I have even sponsored a few conferences, special lectures and the like; I have not, in fine, shown exactly awe-inspiring modesty in my own record of publications and involvements. Not to mention my palpable audacity in initially accepting and then remaining in a very prestigious position at Vanderbilt University Medical Center from 1981 until my retirement in 2002.

Given all this, it must seem odd to others, as it surely does to me, that I would say what I just said about being at times at sea, at others bemused, about this still burgeoning arena.

2.3   A Bit of History

As I've said before, too, I would like very much to help any readers I might have to be just as perplexed as I, for I find myself still astonished at how so many could possibly be so confident about what the field is all about, much less how one actually goes about doing whatever it is that one does as one of its practitioners— I think in particular of becoming seriously involved, as a philosopher, in clinical work. I need to explain why that is, why this venture (and, for some, adventure) remains still so strange, albeit deeply fascinating. It will shortly become clear that my concern centers in particular on what has become known as 'clinical ethics,' although only a very few of its practitioners are actually involved in clinical encounters in the way I and most of my students and colleagues at Vanderbilt were for many years.

Wonder therefore seems to me the proper mood for anyone who chances to look into these matters. But first, along with wonder, a bit of history may be helpful.

What was behind the initial idea, over 40 years ago, of asking philosophers and others[4] in the so-called humanities to come into medical centers (at first, only medical and, a bit later, nursing schools and, much later, into hospitals) in order to participate in medical education and, in some instances, even in clinical situations?

To understand this question is to understand as well a great deal about the subsequent development of bio- (or bio-medical) ethics. It seemed to physicians and medical educators fairly straightforward: considerable help on so-called values and cultural matters was deemed important and it was hoped that the help would come from persons trained in philosophy, ethics in particular.

---

[4] Most often people in the clergy, religious studies, or theology; at times, even some in psychiatry—a phenomenon that has at times seemed to be unusually insightful. More recently, increasing numbers of physicians have come into this field, which most regard as the proper domain only of physicians. This of course ignores that there are many other types of health professional—a point to which many professionals respond by developing their own "expertise" in ethics without, however, solving or even being clearly aware of the still wide divide among different health professional groups and individuals.

With the astonishing new technologies and medical knowledge already at hand in the early 1960s, and even more remarkable prospects on the immediate horizon, (Taylor 1968) physicians had good reason to be troubled (see Chap. 9).[5] Furthermore, new diagnostic tools and techniques promised more accurate, and ever earlier, detection of diseases and anomalies hitherto not available of both present and possible damage—a trend that has continued unabated. Coupled with these were emerging new surgical techniques and instrumentalities, pharmacological interventions, (Farber and Wilson 1961) new types of anesthesia, and all manner of new treatments for conditions not previously treatable as well as for those previously not very effectively treated. Resuscitative techniques along with associated technologies and new medical understanding showed that different body-systems function and cease to function in different ways and paces, and that some could be artificially re-started and supported, thereby allowing needed time for medications to work properly, or healing to take place.

These raised quite awesome, and in some cases wholly new, issues, and gave to many perennial issues a new force and content Gorovitz et al. (1976). Not only was it increasingly possible to maintain patients who only a few years before would have died, often very painfully (as with end-stage renal disease), but also the horizons of life's beginnings and endings were becoming ever more well understood, and though not widely realized at the time they were also being perforce re-defined (Eccles 1970; Penfield 1975; Burnett 1978).

Not that there were no problems; to the contrary. Some perceptive physicians and researchers were already agonizing over the value and moral issues implicit to these developments (Beecher 1959a, b). Recognizing their lack of the training and knowledge to grapple with such issues, they quite naturally turned to others whose credentials at least seemed to bespeak competence, if not expertise (Liddle 1967). Many of these physicians, too, were haunted by the horrors of the Nazi concentration camps, especially the many medical experiments carried on in them, and seemed anxious to realize in practice what was asserted in the Medical Trials at Nuremberg (Howard-Jones 1982; Annas and Grodin 1982; Curran 1982) and by the United Nations Charter, affirming the existence of inalienable human rights, especially for those who are sick, maimed, and vulnerable.[6]

The lingo of the times is suggestive: physicians and others in the so-called health care system expressed (and continue to express) serious dismay over, even while precious little energy has been expended to rectify perceived flaws in, the bureaucratic organization of the modern health science centers and health care more generally, as well as the way new technologies tend, as was often said, "to dehumanize" people. Most health care professionals exhibited genuine concern over the increasing specialization in health care after World War II, which seemed to 'fragment' the 'whole person,' promoting more focus on diseases and organ

---

[5] This mood was accentuated by the late 1970s, with the publication of works by major geneticists.

[6] Although, it must be pointed out, just why vulnerability, illness, and the like, should function so powerfully was not made thematic until much later.

systems than on people (even while every one of them recognizes the remarkable advances achieved precisely by such specialization) (Pellegrino 1979a, b). While many tried earnestly to stay abreast of the ever-growing cornucopia of new developments, substances, sub-specialties and the like, this often meant that physicians were obliged to be and remain technically proficient, and thus that they did not always have the time or inclination to be alert to moral issues, religious values, sensitive caring and other such concerns.

The 'new physician,' avidly discussed as the major part of the agenda for medicine in the 1970s and beyond, it was thought, needed to be 'humanized.' But, as has been pointed out numerous times, it was not in the least clear what this would require nor, in the end, why it was thought to be so important. Such 'humanists,' after all, were hardly the talk of the commons, nor had they made notable or recognized contributions to the common weal, much less to health care. Moreover, to use phrases such as 'medical humanities' seemed only to confuse and bewilder.

One practicing physician, Samuel Martin, lamented publicly in 1972 that it was not at all clear about who was responsible for training that new physician. For many educators, medicine needed to call on humanists—so-called 'experts in human values'—and a new name was quickly concocted for the new breed: *ethicist*, an occupation as unlikely[7] as the name was awkwardly sibilant. Martin and others were dubious about the entire venture. In poignant, if inelegant terms, Martin worried whether so-called humanists "are trying to outscience our scientists. At some time we must deal not only with what makes a humanist, but also with how we can facilitate the transmission of his art" (Martin 1972).

Which could hardly be more to the point: what indeed "makes a humanist," how transmit that "art," and what, in the end, are the "humanities" actually all about? To be sure, Martin perhaps should have worried not only about those trying to "outscience" the scientists, but just as much about those at work cultivating ever sharper and deeper divisions between what C. P. Snow had earlier termed the two cultures. In any event, neither scientists nor physicians, ever more reliant on the biomedical sciences, nor the humanist pretenders to the crown of knowledge were likely to worry about Martin's appeal. Indeed, an appeal to supposed experts in values was not only quite implausible at the time, but for the most part highly improbable given that most so-called humanists were rarely interested in, much less competent to make recommendations about, such matters as were posed daily in the process of creating the new medicine and its supposed new healers: what is death and how ought a person's last days be managed, much less how ought such persons be cared for? For that matter, is a person whose breathing occurs solely because of a ventilator's mournful chug still alive or in some halfway condition never seen

---

[7] After all, it is no mystery that what most believe is the most demandingly practical of human enterprises – clinical medicine – found itself calling on what is, along with poetry, believed to be surely among the most impractical of disciplines, philosophy. The challenge to both still sets the tone for many of their interactions.

before? When does a fetus become a person? Is a comatose individual still a person? The questions only proliferated and, too often, left 'humanists' quite as bewildered as anyone else. Help, in short, seemed ever more distant and unlikely as regards the intense issues faced by health care professionals on the wards, in the nursing homes, or other places where our society tends to house the sick, maim, and elderly.

2.4   The Philosopher's Response to the Physician's Appeal

At first, in the 1960s, only a few, rather venturesome people responded in any way to the appeal in the early 1970s from physicians such as Dr. Martin. Others began to join in within a few years, however.[8] Those who did respond found the world of clinical medicine decidedly eerie, if also compelling. The existential cut of Martin's jibe about what makes a humanist—more to the point, what is a philosopher, what is that art, and how can it be transmitted?—was a keenly felt, daily reality. Separated from comfortable home base in a Department of Philosophy, truth be told, we were utter *naifs* in this strange new world, literal aliens listening in on a esoteric Babel of technical noise and abbreviated, highly technical language.

When in deference to our lack of understanding about how the noise might be (it wasn't always) translated into English ("What does 'PTA' mean?" "Oh, that's 'prior to admission'!"), our ignorance only became all the more plainly pathetic, as we were sometimes stunned into silence ("PDA?" "Oh, that's 'patent ductus arteriosis,' which if we can't do anything about it, means certain death for this baby!"). When we were nonetheless encouraged to talk or offer some opinion ("...whether a child born with developing hydrocephalus secondary to myelomeningocele should have a shunt instituted..."), we found ourselves babbling in an equally alien tongue about moral agents, persons and potential persons who could in all likelihood never become persons, but who yet perhaps should, that is, might be treated as if they were persons or as if they still had or might be said to now possess some kind of moral status. . . . Not particularly helpful in the press of the circumstances which occasioned the request in the first place.

Many philosophers recoiled in shock and dismay: *this* simply is no place for a philosopher: our education, training and disposition includes nothing that could in any way prepare us for rendering such judgments, much less making definitive moral declarations on the spur of such critical care moments. Indeed, even if one could, *per impossible*, begin to untangle some of the densely packed moral themes and issues presented in what were termed 'clinical cases,' we could only lament that we had neither the time to do so properly nor the appropriately prepared audience to

---

[8] My appointment in 1971, first on a grant then, my second year, on the medical faculty at SUNY-Stony Brook was the first ever line-item for a philosopher in medicine in the State University of New York's official budget.

listen to and hear the discourse, much less participate in subsequent indispensable philosophical discussion.

Nor did gradual familiarity with clinical settings, specific types of conditions, situations and patients, technical jargon, exotic technologies, and the rest help to ease the razor-sharp sense many of us felt: that the philosopher remains an inter-loper, a stranger in a strange land, a theorist in the land of therapists. The philoso-pher's stock-in-trade, I was reminded more than once (by colleagues as much as by physicians), is principles and norms (remember Hume, Mill, or Kant!), and neither therapy nor guidance counseling. The philosopher's business is foundations, ideas, concepts, and logic, not, as often seemed to be the real agenda, sensitizing health professionals to value phenomena, nor engaging in acts of persuasion to try and convince them to act more humanely. The mind is to be *studied*, not expanded, by our labors! Even so, the movement got quickly underway, and already showed remark-able growth during the 1970s and early 1980s (Pellegrino and McElhiney 1981).[9]

What rapidly developed, especially during the latter period, is readily under-standable, even if, as I'll suggest, somewhat dubious. The new arena of concerns (and, truth be told, of employment[10]) began to be viewed simply as a different place to conduct the usual sort of business of philosophers: writing scholarly tracts, talking with each other, and teaching courses (with appropriate modifications to accommodate the intensely practice, problem-oriented, and professionally moti-vated students of medicine).

Not only did this turn to accustomed pedagogy tend to dull the knife-edged issues occasioned by clinical situations; it was also widely urged that such encounters were quite unnecessary and possibly even obstacles to the conduct of sound philosophy. Thus, many agreed with the notions that the philosopher is simply out of place in clinical settings, and that physicians are seriously misled if they look to philosophers for solutions to the questions of human conduct and decision faced by physicians (only rarely were patients and families, much less nurses or other health professionals mentioned in these contexts). While medicine was seen as presenting fascinating and even demanding social and moral issues, it was generally assumed that philosophers could properly address them solely in philosophy's usual ways (Shaffer 1975).

Ethics, for all its traditional emphasis on practical reason, was typically regarded by almost everyone as a *theoretical*, not a practical, much less therapeutic disci-pline. Medicine was thus quickly, and with noticeable relief, interpreted as merely one among many of the 'fields' to which philosophy was to be 'applied,' through its

---

[9] In the sole study of this growth at the time, in which I participated, by the Institute on Human Values in Medicine, it was noted that from meager beginnings in the early 1960s, to a bare handful of programs nationwide when I became involved in 1971, the movement had become a true movement by the early 1980s: by then, all but one or two medical schools in this country and most in Canada and Australia had included some form of 'human values' training, some of it very questionable; and by the mid-1980s, the field had grown to international proportions.

[10] Of no small concern to graduate programs in Departments of Philosophy, not to say the American Philosophical Association, at a time of serious retrenchment by universities, hence, of fewer and fewer positions.

familiar advocacy of one or another set of ethical principles, analysis, and argumentation. Not surprisingly, biomedical ethics swiftly became known as 'applied ethics'—not unlike the engineer who 'applies,' say, the rules and notions of structural engineering and, more basically, of physics—a view that has to my knowledge rarely been seriously questioned. The ethical analyst's task was to study such knotty words as 'good,' 'evil,' 'right,' 'wrong,' 'decision,' 'responsibility,' 'action,' and suchlike; nothing more. In a rare moment of candor, R. M. Hare openly declared:

> Philosophy is a training in the study of such tricky words and their logical properties, in order to establish canons of valid argument or reasoning, and so enable people who have mastered it to avoid errors in reasoning, and so answer their moral questioning with their eyes open. It is my belief that, once the issues are thoroughly clarified in this way, the problems will not seem so perplexing as they did at first. . . . (Hare 1977)

While it might be prudent for such a philosopher to make periodic forays into clinical life—the analogy usually appealed to was physics: that the philosopher could learn much by being in proximity to physicists—even, it may be, to meet with a patient or two, this is not in the least necessary nor relevant, and could be a positive hindrance to his or her proper analytic task.

Governed by the idea of 'application'[11]—'applying' ethical rules, norms and principles to practical problems—there grew up the familiar range of articles, books, anthologies, and of course, textbooks. First typically presented in the latter was a familiar menu of moral 'theories'—deontology, utilitarianism, natural-law, virtue-ethics, and their many variations—usually coupled with a more of less harsh glance at medical oaths and codes to demonstrate their woeful inadequacy (presumably because they were typically regarded as terribly un-philosophical). Then there followed the also familiar litany of supposedly obvious 'moral problems:' abortion, euthanasia, damaged neonates, scarce resources, human experimentation, and the like. The idea was, having grasped something of the available theoretical equipment and alternatives in ethics, to show then how each is or might be 'applied' (through text or collected articles) to that range of practical problems, and then to suggest the usual, also typical or standard difficulties each faces in being thus 'applied'—leaving the rest of the work, presumably, up to the physician's choice: choose which of the items on the menu best suits your needs, then go for it in the way supposedly laid out in the text.

By that point, too, of course, biomedical ethics seemed thoroughly domesticated into the usual packages of concepts and methods, courses and conferences, speeches and articles. The bite of medicine's initial appeal was then swiftly co-opted by official philosophy in much the way as the hippie's tattered blue jeans were by the fashion world.

---

[11] The term still niggles with ambiguity: just what could it possibly mean, in actual and concrete terms, to 'apply' a 'rule,' a 'principle,' or, say, Kant's 'moral law?' The sense of this act was never submitted to careful scrutiny.

2.5 How Did Physicians Respond to This Response?

The idea that medical professionals needed to be 'ethical' had not yet quite caught hold; indeed, it remained quite unclear, although it was in the making. Yet, few were the physicians bold enough to say harsh things about ethics, though there were some, and their voices were significant. Again, a bit of history is instructive.

In the mid-1970s there occurred a serious backlash against ethics, as it was termed by one of the major commentators of the time, Daniel Callahan (1975, p. 18). More to the point, while some physicians, like Alan R. Fleishman, helped to initiate programs in what was termed "ethical analysis" for residents and medical students, reports of this labor tended to be monuments of double-talk. For instance, Fleishman stated in one report on his program that while resident physicians presumably learned from his program that "their decisions were based on ethical principles," they yet uniformly "felt that the neonatal ethics rounds did not specifically affect medical care." Some of the "most frequently presented issues involved the rights of the fetus and of the newborn," over against the "right to decide" of the parents. Yet just these issues and the "principles" supposedly "applied" to such situations were regularly "found to have little relevance in actually determining what was the right decision." Moreover, while residents affirmed that "they did increase their understanding of ethical principles and the process of ethical analysis," they also stated that "they felt their general moral and ethical views had not been changed" by the program. Nevertheless, with unnoticed irony, the program continued (Fleishman 1981).

The message had to be obvious to any physician: although expressed in glowing terms and recommended to other medical units, the program was just as obviously a complete failure—as the residents clearly recognized. It changed no one's behavior, decisions, or moral views. What the 'ethicists' did and said had no relevance to clinical judgments. Yet, the "moral conflicts," it was alleged, which regularly occurred in that neonatal unit, were supposedly "handled" by the "process of ethical analysis"—a term, it must be noted, that received no comment whatever, much less clear explanation of what was actually done under its aegis. Beyond all that, one can only wonder (since it was not part of the report) what the parents of those babies, or, say, the nurses who regularly tended to these babies, thought about this program— in the unlikely event that they were actually informed or made aware of it.

The message here shouldn't be lost—for instance, blaming the 'ethicist.' Other physicians frequently reported precisely that kind of dismay. One physician seriously involved with philosophers for some time, and in fact the instigator of one of the major ethics programs in the country,[12] Eric Siegler, was acutely disillusioned over what he termed "the biomedical ethics establishment (BME)." He argued that

---

[12] He established the Center for Biomedical Clinical Ethics at the University of Chicago, a still-ongoing concern.

those involved in BME were too inexperienced and insensitive to the routines and rigors of clinical practice, and in fact tended to be "hypercritical" of the Hippocratic tradition and commitments of physicians. With others, he lamented the proliferation of BME teaching (which he thought took up issues quite different from those encountered by real clinicians), and the virtual dominance of the field by non-physicians. Thus, he argued, not only do those within BME have quite different agendas from physicians. Such philosophers and others in BME also are in the end merely observers who exhibit little more than the "counterfeit courage of the non-combatant;" while physicians, on the other hand, are legally, morally, and professionally accountable to their patients, philosophers are not. "Philosophers," Siegler insisted, "are theorists with no need to come to conclusions about specific patients or cases," while physicians "must constantly deal with specific cases, decision-making, best guesses, and directed therapy" (Siegler 1979).

Accordingly, Siegler, Fleishman and others urged that physicians must counter-act the BME establishment: even more, physicians themselves must become expert in ethics. After all, only physicians[13] are experienced and can be knowledgeable therapists, only physicians are held accountable for any- and everything they do, and only physicians know the uncertainties and terrors of actual clinical practice: life in the trenches, so to say, is open only to physicians.[14] Siegler and others expressed gratitude to those in the BME, even to that "establishment" itself (whatever it may have been); but physicians had to be deeply skeptical about the supposed fruits of the almost four-decade-long effort of serious flirtations between medicine and philosophy. Richard M. Hare's unwitting admission might have come home to roost if only philosophers had heeded what seems more warning than not, despite his views (above) on matters of philosophical ethics:

> I should like to say at once that if the moral philosopher *cannot* help with the problems of medical ethics, he ought to shut up shop. The problems of medical ethics are so typical of the moral problems that moral philosophy is supposed to be able to help with, that a failure here would be a sign either of the uselessness of the discipline or of the incompetence of the particular practitioner. (Hare 1977)

I take it that, of course, no philosopher, Hare especially, would in the least concede to having to close up shop. The rejoinder to physicians such as Siegler, while it surely gave no comfort, may yet have pacified philosophers: what physicians were initially asking was just a plain mistake—understandable, perhaps, for they are not philosophers. What philosophy can do, and do quite well, is study and clear up the underbrush of the tricky words of moral discourse for their logical properties, cultivate respect for the canons of clarity and valid argument, provide distinctions between fact and value, medically descriptive and evaluative factors,

---

[13] Not atypically, Siegler almost never mentions nurses and other health professionals, not to mention patients, families, or close friends.

[14] Again, mention of patients, parents, families or their close friends is curiously absent, along the prominent absence of other health professionals, nurses in particular.

and suggest ways by which moral theories, principles, axioms and rules should be 'applied' to practical clinical problems. To expect more is to ask that philosophers go beyond their proper place and competence—which could only erode if not destroy the integrity of philosophy itself, as would surely occur to medicine were physicians invited to practice medicine in a department of philosophy.

To be sure, were these the final words, it is obvious that the very idea of 'application,' especially 'applied ethics,' would thereby become incoherent: patent nonsense that only re-establishes the sharp and always divisive line between the two cultures.

2.6   Brief Overview of Medicine

A brief word about medicine and its history seems also helpful to understanding many of the complexities faced by those of us who were once invited into this new kind of endeavor. This will, I hope, begin to clarify why I do not agree with Hare, Fleischman, or Siegler. Eventually, this will make it necessary for me to lay out, explain and defend an understanding of philosophy and ethics that is quite different from the view underlying what they have said. As for medicine, all I can do is present what I have come to understand about this enterprise—which begins with what follows.

In general, medicine is fundamentally a teleological discipline; that is, it is oriented toward goals or ends, and it is in these goals that the distinctive features of medicine are clearest. Three of these are central to the enterprise in all its expressions. First, it is a therapeutic, hence preeminently practical discipline, in that it is oriented towards healing, seeking to restore, normalize, or at least ameliorate the effects of illness, injury or the consequences of genetic and/or congenital error and/or defects. This is expressed most clearly in the best-known parts of the Hippocratic Oath: always to act "on behalf of" the sick and/or injured person. It is this characteristic that gives knowledge its origins and purpose: it is in essence defined by its orientation towards helping, healing, ameliorating. More on this will of course be necessary to provide at a later point.

Second, and the other side, so to speak, of the first is that its practitioners are enjoined not only to help, but "first" of all "to do no harm" nor cause "mischief" to persons who come to them for help. They are directed to act so as to prevent those who come for help from doing "harm" or "mischief" to themselves. In this sense, each healer who takes this oath is focused not only on helping and not harming the patient but also on disciplining him/herself, and patients: the healer is enjoined to restrain him/herself from causing harm, but also to keep patients from harming themselves.

Third, healers are charged to make every effort to prevent, impede, and even to try and stop the negative effects of disease, a goal that is accomplished in part by promoting measure in one's own and in patients' lives, especially by means of controlling or at least advising on proper bodily intakes: *dietetics* is as vital to the tradition as is diagnosis and therapy. Medicine is thus oriented at one and the same time toward restoration (diagnosis, therapy), restraint (on self, on patients), and improvement (dietetics)—and in this sense, it is directed not only toward therapies but also toward melioration and eugenics. These themes need more careful examination.

To be sure, traditional Hippocratic medicine shows a number of other significant themes. I think particularly of its several interpretive schemes for understanding and explaining "symptoms"—crucial as the basis for therapy; the place of and rationale for anatomy; or the essential asymmetry of the clinical encounter. Here, however, I want to attend, albeit very briefly, to only several of medicine's orientations.

Within this tradition is a clear emphasis on detecting and treating disruptions of the human bodily organs, tissues, or structures, whatever their source—illness, injury, environs, past experience, diet, or the debilitating circumstances of birth and/or genetic heritage. There are, of course, several quite different approaches to both detection and therapy, and these show clear differences—from the Dogmatic tradition's conceptual commitment to disease as pathologies considered to be internal to the body with symptoms of these *dyscrasias* appearing externally, to the early Empiric and later Methodist tradition's idea that bodily symptoms indicate past, historical experiences in one or another sort of environs or due to previous dietetic intakes. Coordinate with the first was a form of reasoning termed *analogismos*: going from external 'effects' to internal 'causes.' The latter, on the other hand, embodied another form of reasoning termed *semiosis*: proceeding from present symptoms to past occurrences and/or intakes. Similar differences can be found among the various forms of therapy—from the notion that only substances that are 'unlike' the disruptions constitute successful therapy, to the opposite idea that only substances that are 'like' the disruptions are appropriate (Zaner 1992).

In each case, and from any of the several sub-traditions that subsequently arose, one basic goal was to work toward the *restoration* or amelioration of disrupted bodily functions. The second goal, equally basic, is closely tied to the idea that whatever the healer also does (for instance, anatomical dissections), everything learned is strictly in the service of each individual patient who comes for help. Therapy is thus the central goal of medical attention; indeed, even the pertinent knowledge the healer acquires in this process of detection and treatment "is discovered in medical practice itself and is derived solely therefrom" (Edelstein 1967, p. 201).[15] While there surely are forms of knowledge not so tightly connected to therapy, to practice, that which defines the medical healer is both discovered from and devoted to helping patients. As Jonas emphasized, practice is not incidental, but essential, to science and, I've urged, to medicine.

---

[15] Indeed, Edelstein suggests, this emphasis on the fundamental place of encounters with the patient is, he says, "medicine's own creation and, it seems to me, its original contribution." Then, in a footnote to this sentence, he emphasizes, "if the Greeks have a dislike for the individual and a preference for the typical, the counter-balance is provided by medicine, not by geography, history, or another science" (note p. 18).

Thus, the well-known ancient Greek preference for the typical, the similar, and the universal in its main conception of knowledge, is not found in medicine whose emphasis in on *praxis*, on the goal of helping the unique and individual. It is not so much what is 'similar' or 'the same' among different patients; instead, it is what is *different* or *unique* that is the main interest—an emphasis which stems directly from medicine's strict focus on diagnosis and treatment of the individual person. For just this reason, not only is treatment of individual sick or injured persons a central point of restorative medicine, but the healer made "the afflictions of the human body the law governing his treatment," in Edelstein's words: disruptions of or in the body inform the healer what should be done.

Closely connected with acting always "on behalf of" each individual (the historical core of the idea of benefit), on the other hand, is the phenomenon of *mischief*. An intriguing twist, noted already, is included, however. In what I have termed its second goal, the healer is severely admonished to remain "free of all intentional injustice, of all mischief," such as having sexual relations with any patient. But equally, the healer is also cautioned at all times "to keep them [that is, patients, families, households] from" doing harm, injustice, or mischief to themselves.

There are two facets of what I might term the "mischief" thesis (Edelstein 1967, pp. 39, 53).[16] First, each bodily appetite is regarded as an inclination and even craving of the soul and, whether acquired or native, such appetites tend of themselves toward indefinite increase if they are not somehow held in checks. Such unchecked increase constitutes an indulgence, furthermore, and indulgence leads to unhealthiness and disease. Therefore, the quality and quantity of all nourishment must be chosen with great caution, a talent or ability that is the "supreme wisdom entrusted to the physician" (Edelstein 1967, p. 24)—just this constitutes the core of medical knowledge. Dietetics is a veritable *discipline* learned through the rigors of medical experience, and constitutes the principal avenue through which the native tendency of any appetite toward increase can be controlled. Dietetics was thus among the primary therapeutic measures available to the healer.

The second facet of the mischief thesis is that the abuse of a proper dietary regimen is an *acquired* inclination. Accordingly, there is a necessary *moral* element at the heart of the dietetic discipline. Since Hippocratic healers thought that people are for the most part unable to learn this discipline on their own, they must be instructed and guided by the healer. The moral element here has two sides: healers are obliged to keep people free from self-abuses, on the one hand, and on the other, in order to attain and remain in a state of health, people generally are enjoined to avoid over-indulgence and for this must comply with the healer's directives—

---

[16] Edelstein's close analysis of the Oath convincingly shows that, written during the fourth century B.C., it is an expression "of Pythagorean teaching...thoroughly saturated with Pythagorean philosophy...[and] is a Pythagorean manifesto." And this provides the clue to understanding what the Oath cautions against in this passage. The healer pledges to "guard his patients against the evil which they may suffer through themselves," for it is a major thesis of the Pythagoreans that "men by nature are liable to inflict upon themselves injustice and mischief" (pp. 22–23).

which stem from the healer's special knowledge about the appetites. To act on behalf of the sick and injured (as well as their families and households) is in part to keep them from harming themselves: treatment of disease is therefore not the only theme of the physician-patient relation, since the physician is also charged with dietetic instruction, which is suffused with moral elements.

To appreciate these, it seems to me helpful to pause in order to probe one of the more pervasive themes in Greek culture.

2.7 An Interlude: Gyges, Aesculapius and the History of Medicine

Even though medical therapies were for many centuries virtually useless, frequently painful, and medical understanding often erroneous, the relationship's asymmetry of power has been a fundamental and remarkably unaltered component of medicine's self-understanding almost from its inception. Rarely understood in this way, this is nevertheless an essential component of restorative medicine in the Hippocratic tradition. Steeped in an understanding that the relation to patients must be governed by certain fundamental virtues prompted by that asymmetry, these ancient physicians had a remarkable insight, I believe, into a key facet of the moral order that governs, as they understood these matters, the relationship with patients, families, and households.

Ludwig Edelstein's lucid studies show that the principal generic virtues in the Hippocratic Oath are justice (*dike*) and self-restraint (*sophrōsyne*). Whatever one may think of the at times barbaric treatments practiced under medicine's aegis (almost into the twentieth century),[17] ancient physicians realized full well that medical practice involved the physician in the most intimate kind of contact with variously compromised and vulnerable human beings—and required decisions that would invariably affect the patient and family, sometimes in profound ways. On the other hand, it was also realized that the patient faced an urgent issue: "How can he be sure that he may have trust in the doctor, not only in his knowledge, but also in the man himself?" (Edelstein 1967, p. 329). This critical blend of virtues—most accurately, perhaps, judicious restraint—was, not unsurprisingly, regarded in the ancient Hippocratic texts as the primary sense of medical wisdom (*On Decorum* and *On the Physician*) (Edelstein 1967, pp. 6–35).

To be a patient is to be intimately exposed and directly vulnerable to actions of others who claim to be capable of healing actions (even while doing that may cause harm: at times the only avenue for healing is to cause harm, as these ancients well

---

[17] Because almost any therapy harbors pain and disruption of its own kind, this characteristic persuades me that courage should surely be included as equally basic, for both patients and physicians (who, ordained to "help" and "not cause harm," must at times nevertheless precisely do that as the way to provide that help).

understood).[18] Because of the specific type of knowledge unique to medicine, the healer's possession of drugs and technical skills, and the access to the intimate spheres of patient life, the healer clearly realized that they were in a unique position *to take advantage* of patients while, by contrast, patients were *disadvantaged* both by illness or injury and by the very asymmetry of the relationship with some healer. Precisely this appreciation of the asymmetry of power in favor of the physician led to an understanding of 'the art' as a fundamentally moral enterprise under the guidance of central virtues: justice and restraint (to which, as noted, courage must be added).

A hospitalized patient once poignantly remarked, "you have to trust these people, the physicians, like you do God. You're all in their hands, and if they don't take care of you, who's going to?" (Hardy 1978, p. 40). Noting how "overpowering" doctors can be, another emphasized, "They've got an edge on you" (Hardy 1978, pp. 92–3). In these plaintive words is the echo of an ancient puzzle—the temptation ingredient to having actual power over the existentially vulnerable patient—all the more keenly ironic within medicine, supposedly governed by the Oath, almost unchanged throughout its long history.

This puzzle, I am convinced, is at the heart of the Hippocratic tradition in medicine, especially of the virtues long regarded as fundamental to it. It is especially plain in light of the Oath's apparent mythic sources with the god Apollo and his progeny, Aesculapius, "the god of doctors and of patients" (Edelstein 1967, p. 225). Physicians who took the Oath regarded themselves as bound by a covenant to help sick and injured people of all sorts, without bias. In order to help, they became involved with vulnerable people in the most potent and intimate ways, at times called on to render judgments and make decisions that reached far beyond the application of merely technical knowledge and skills. They thus believed they were entrusted by the gods with a supreme wisdom about afflicted people, committed to be "physicians of the soul no less than of the body" (Edelstein 1967, pp. 24–5).

The Aesculapian healing-places were open to every sick or injured person, whether slave or free, pauper or prince, child or adult, man or woman. Following the guidance of this "god who prided himself most of all on his virtue of philanthropy," the healer understood that he thereby took on certain fundamental responsibilities. Sarapion laid out some of these in a poem inscribed on stone in the Athenian temple of Aesculapius: "First to heal his mind and to give assistance to himself before giving it to anyone," and only then to "cure with moral courage and with the proper moral attitude...For we are all brothers" (Edelstein 1967, p. 344).

The covenant incorporates an understanding of social life—including what brought the vulnerable sick person face to face with the healer and powers of the 'art.' The Oath's covenant invokes a moral vision at the heart of the healer-patient relationship, and shows a strong sense of the power inherent in the art, of the potential for control and even violence to the patient who was "all in the hands" of

---

[18] A point that, we will see later, constitutes the ironic core of "trust," the necessity of which is at the heart of the clinical encounter.

the physician. Acting "on behalf of" the sick person and maintaining strict "silence" are as integral to the Oath as certain conducts were strictly banned.[19] It thus incorporates that peculiar blend of justice and restraint (and courage) to govern the relationship. This implies that physicians evidently recognized that they were in a unique position to take advantage of people when they are most vulnerable and accessible. It also strongly suggests recognition of the central challenge and temptation inherent to the work of physicians, thus demonstrating the emergence of a sophisticated moral cognizance (Zaner 1988/2002, pp. 202–23). *Vulnerability*— and its correlates, personal *integrity* and *dignity*—were therefore just as essential to the art as was acting on behalf of and never harming the patient (beneficence and non-malificence)—indeed, the moral force of the latter is found in that vulnerability.

Edelstein emphasizes that the Oath lays out for this "sacred art" a "morality of the highest order;" the healer was enjoined to "a life almost saintly and bound by the strictest rules of purity and holiness" (Edelstein 1967, pp. 326–7). To practice medicine was and is deliberately and voluntarily to assume the responsibility for being morally attentive—responsible for and responsive to each and every individual who seeks aid, and to accept the bond of a covenant with each person, but also with his or her family and household.

The moral cognizance at the heart of the Oath is striking but, as noted, forces a searching moral question: What could possibly move any physician *not* to take advantage of the vulnerable patient? Why not take advantage, especially when the patient is, precisely, vulnerable? Buried squarely within the Hippocratic tradition, is that ancient puzzle. One need only consider another, equally ancient and powerful myth about the temptation of having actual power, to put the puzzle into perspective: the legend of the Ring of Gyges in the Second Book of Plato's *The Republic*.

Having gained the power of the ring (to become invisible when the ring's collet is turned) found in the belly of a bronze horse (uncovered in a crevice by an earthquake), Gyges is then able to do whatever he wishes. And, he does just that: seduces the queen and, with her by his side, slays the king and becomes king himself. The puzzle within the Hippocratic Oath is striking: *having* the advantage over the other, the power, a physician persuaded by Gyges would surely *take* advantage, just because, given the ring and its power, the patient is vulnerable and readily accessible (as were the queen and king of Lydia). Interpreting medicine from the perspective of the social milieu postulated in the Gygean myth, the Oath itself is either patent nonsense or a mere façade for the exercise of power. With that, moreover, is an implicit degradation of the moral sense: it easily becomes a sort of "merely moral," a mere aid for the pursuit of power.

When people are strangers, there is all the more reason for suspicion and distrust as the basic form of social orientation, since the very grounds for trust in the helping relation are missing, or at the very least are quite problematic. For among strangers, on the one hand, there is nothing common, enduring, and mutual to prompt or

---

[19] For instance, abortions and providing lethal substances for suicides.

sustain understanding: neither the healer nor the one seeking help knows what, if any, values they share nor how their respective values differ. Is the healer trustworthy? Does the patient mean what she says? Just so, at the core of the clinical event is the asymmetry of power in favor of the healer over against the vulnerability of the one seeking help from the healer. While the healer has the power to influence the patient, often without his/her knowing, the healer doesn't know how this power is regarded by the patient; but neither does the healer know whether she or he is trusted to use power for the patient's benefit.

But if the Gyges myth is alien to what I've suggested about the Hippocratic understanding of medicine, it nevertheless provides a significant highlight on the key moral issue—the power over vulnerable others, which highlights the moral place of dignity and integrity, and gives a cutting edge to the central bioethical issues of our times (Kass 1977).[20] If healers are to be entrusted with such power and intimacies, the crucial question concerns what they must do and be to deserve that trust (that is, to be trustworthy). Why *not* use the power of that asymmetric relation for the healer's own advantage? The patient must trust that the advantage will not be taken, abuse will not be done while yet being uniquely at the mercy of the physician—the very one who professes and then proceeds to use the power of the art (knowledge, skills, resources, etc.), who proposes and then proceeds to engage in highly intimate and consequential actions on people and their circle of intimates at the very time when they are most vulnerable, at times bringing about important forms of change in what, how, even who, they are or hope to be.

These myths invoke deeply different and conflicting visions of the social order, especially the social context of clinical encounters. In both, one with power confronts another at an intrinsic disadvantage. For the Hippocratic, the potencies of the art were clearly appreciated and given recognition and moral expression in its Oath: the injunctions to act always "on behalf of" the sick person, never to take advantage of the patient or his family/household, never to "spread abroad" what is learned in the privacy of the relationship with the sick person. For the Gygean, however, the therapeutic act can make no sense: why engage in helping, after all, since that would merely not only permit but actually assist the vulnerable person to become less vulnerable, less open to coercion?

But even as *therapeia* was understood as Hippocratic-Aesculapian, the grave moral issues in no way disappeared but instead became an abiding part of medicine's history: why "first do no harm" then "act in the patient's interest?" The response to this question has typically suggested that, whatever else medicine is or does, it is in the first instance devoted to helping those who are sick or otherwise debilitated—helping them, that is, providing therapies designed ultimately to restore their damaged bodily functions or abilities and thereby rectify that otherwise inherent asymmetry of power.

---

[20] It is, for instance, a principal argument used by those opposed to certain medical procedures: abortion, in vitro fertilization and, even more, stem cell research and human cloning.

2.8 Medicine's Clinical Practice

Therapeutic theories in all their variety are attempts to make sense "of the healer's experience with the patient," (Coulter I, 1975, p. viii) in order then to provide some substance or procedure to help the person afflicted. In different terms, to think about the clinical event is to discover "the universal fact that humans become ill and in that state seek and need help" (Pellegrino 1983, p. 162). No matter how ill, the ill person presents in a

> special state of vulnerability and wounded humanity not shared by other states of human deprivation and vulnerability... In no other deprivation is the dissolution of the person so intimate that it impairs the capacity to deal with all other deprivations. (Pellegrino 1982, p. 159)

A disturbance within that most intimate sphere of relationships between self and its own body, vulnerability and its coordinate appeal for a healer to help, arises as an appeal "to restore wholeness or, if this is not possible, to assist in striking some new balance between what the body imposes and the self aspires to" (Pellegrino 1983, p. 163). If full restoration is not possible, then amelioration, adaptation or coping, palliation, etc., become the ends of the healing relation.

These ends are specific to 'the art' and distinguish it from other human activities, as well as from other activities in which healers may also engage. For instance, when the causes and pathogenic mechanisms of disease are sought in the form of a biomedical experiment, the end is primarily knowledge. The end of preventive medicine, on the other hand, is mainly to preserve the well-being of individuals and groups, whereas social medicine seeks the health of an entire population, the public good. Healers are important for all these, and while the clinical skills usually associated with being a healer surely are important for those other tasks, only within the healing relationship itself is the healer required to brings those skills directly to bear.

The defining moment of medicine is, then, the clinical event, and it is within this direct and always intimate relation with the vulnerable ill person that the healer seeks actions that are both technically right (scientifically sound) and, with the patient (and loved ones), morally good (a healing action). To provide that help, certain questions must be answered with each patient: What is wrong, what has it done, and how will it affect the patient? What can be done to help? What *should* be done? "These converge on the choice of an action that is right and good." Thus, the moment of decision of each healing relation is the centerpiece and "true clinical moment of truth, and in that moment what is most characteristic of medicine comes into existence" (Pellegrino 1983, p. 183).

Thus, to view medicine merely or mainly as a matter of knowledge is critically inadequate: merely to possess biomedical knowledge does not imply that the patient, and what healing is, are understood, much less that the patient's interests will be served, error appreciated and discussed, etc. Rather, oriented towards therapy—medical knowledge is inseparable from its *praxis* (Jonas 1966, pp. 194–5)[21]—medical knowledge is essentially directed to and governed by the relationship to the one who is ill, vulnerable, and anxiously appeals for help from the person who has or claims to have knowledge and an ability to help. Knowledge is first of all in the service of action, specifically here the interests and needs of those who are ill or otherwise disabled. Hence, clinical medicine is to be understood not by knowledge of itself, but rather as the clinical event, the moment of clinical truth.

There is an inherent logical difficulty here: each patient is more than merely an instance of some scientific principle or statistical norm, even while such principles and norms are surely pertinent as regards the patient's condition. In somewhat different terms, understanding the biology of disease requires that disease symptoms and their sundry mechanisms be abstracted from individual patients then generalized into commonly recognizable diagnostic disease patterns (which process in ancient medicine was termed the "logical classification of diseases"). Diseases are typically expressed in fairly constant ways in cells, organs, or enzyme systems; similarly, a person's genetic makeup or changes in the immune system can alter his or her biological reaction to diseases. As is suggested by clinical interventions, however, it is equally clear that personal habits, environs, diet, physical conditioning, and the like can also alter that reaction. Each illness, Cassell says, "is unique and differs from every other illness episode because of the person in whom it occurs. Even when a disease recurs in the same individual, the illness is changed by the fact that it is a recurrence... [T]he presentation, course, and outcome of a disease can also be affected by whether the patient likes or fears physicians," for instance, as well as other factors unique to each patient. These concerns have become all the more critical in light of the problems presented by the major diseases of our times: heart disease, cancer, stroke, ulcers, diabetes, even the malignancies associated with AIDS, which stem "primarily from the way we live" (Cassell 1979/ 1985, p. 16). Thus, treating them requires sensitivity to these modes of actual living, quite as much as do the range of chronic illnesses.

Uncertainty can be reduced to some degree by having the best available information at hand and insuring that meticulous attention is given to the clinical arts: history taking, physical exams, critical use of probabilistic and modal logic, and mastery of the art of clinical listening and dialogue. It is equally and morally imperative for the healer to understand what illness means for each patient. Illness

---

[21] As mentioned earlier, Hans Jonas has emphasized that the practical use of scientific knowledge is by no means accidental to modern science and theory more generally—and, it seems perfectly clear, to medicine as well to the very extent that it has allied itself with that science. Theory and power are integral to one another, he long ago argued: "the fusion of theory and practice becomes inseparable in way which the mere terms 'pure' and 'applied' science fail to convey. Effecting changes in nature as a means and as a result of knowing it are inextricably interlocked." Hence science is technological by its nature. At the same time, to very extent that medicine's theory and practice is ordained to the diagnosis, therapeutic assessment, and prognosis of specific patients, it is a matter of practice as well. Precisely this characteristic was noted by Edelstein in his studies of classical Greek Methodism: this understanding of the close alliance between theory and practice— that practice informs and shapes theory—"is medicine's own creation and...its original contribution" (Edelstein 1967, p. 201, n18).

is an experience that challenges, often in a critical and deeply personal way, the meanings of personal life, suffering, relationships with others, and with the person's fundamental values. Thus, it is imperative for the healer to elicit, listen for, understand and be understanding of the patient's own experiences and understandings of his/her presented illness.

2.9 Illness and Disease

Taking patients' histories and engaging in clinical conversations with them and their families (or whomever they include in their circle of intimates) have thus become increasingly recognized as central elements of the clinical event. Beyond efforts to determine 'what's wrong' (the diagnostic moment), such conversations are typically aimed at laying out and understanding available therapies for the patient's specific problems, then jointly devising strategies of intervention with patients and/or loved ones. These strategies are necessary, not only because their effective realization depends on the patient's initiative, compliance, and discipline (as well as support and understanding of family and/or significant others)—precisely why courage is a vital virtue in these situations. Beyond this, it has become well recognized that the patient or legal surrogate is the real authority for decisions—arising mainly from situations involving the initiation or withdrawal of life supports at the end of life.

It is crucial, Norman Cousins has argued, for healers to learn to "strike a sensible balance between psychological and biologic factors in the understanding and management of disease" (Cousins 1988). Personal and emotional life has, he insists, too long been regarded merely as "intangibles and imponderables." Instead, there is a "presiding fact" in these inquiries: "namely, the physician has a prime resource at his disposal in the form of the patient's own apothecary, especially when combined with the prescription pad" (Cousins 1988, p. 1611). Or, in Cassell's words, "the illness the patient brings to the physician arises from the interaction between the biological entity that is the disease and the person of the patient, all occurring within a specific context" (Cassell 1985, pp. 4–5).

Patients organize and embody the illness experience most often in narrative formats, deeply personal though often truncated stories; (Frank 1991, 1995) it is thus imperative for healers not only to recognize each patient's story, but also to develop and refine their abilities to talk about themselves, to encourage voicing and, eventually, interpreting their stories. Frequently, however, neither the patient nor family is able to express their full narrative adequately or accurately—surely a requirement for judging whether s/he is truly informed, uncoerced, and capable of making decisions. These considerations lead to several points bearing directly on disciplining the healer's interpretive intelligence in clinical conversations. These will be taken up in subsequent chapters.

References

Annas, George J., and Michael A. Grodin (eds.). 1982. *The Nazi doctors and the Nuremberg code*. New York/Oxford: Oxford University Press.

Beauchamp, Tom, and James Childress. 1983/2012. *Principles of biomedical ethics*. New York: Oxford University Press.

Beecher, Henry. 1959a. *Experimentation in man*. Springfield: Charles C. Thomas.

Beecher, Henry. 1959b. Experimentation in man. *Journal of the American Medical Association* 169: 461–478.

Burnett, Sir Macfarlane. 1978. *Endurance of life*. Cambridge: Cambridge University Press.

Callahan, Dan. 1975. Scannings: The ethics backlash. *The Hastings Center Report* 5(4): 18.

Cassell, Eric J. 1979/1985. *The healer's art: A new approach to the doctor-patient relationship*. Cambridge, MA: MIT Press.

Cassell, Eric J. 1985. *Talking with patients*, vol. 2. Cambridge, MA: MIT Press.

Coulter, H.B. I: 1975. *The divided legacy*, 3 vols. Washington, DC: Wehawken Book Co.

Cousins, Norman. 1988. Intangibles in medicine: An attempt at a balancing perspective. *Journal of the American Medical Association* 260(11, September 16): 1610–1612.

Curran, William. 1982. Subject consent requirements in clinical research: An international perspective for industrial and developing countries. In *Human experimentation and medical ethics*, ed. Zbigniew Bankowski and Norman Howard-Jones. Geneva: Council for International Organizations of Medical Sciences.

Eccles, John C. 1970. *Facing reality*. Heidelberg/Berlin: Heidelberg Science Library/Springer.

Eccles, Sir John. 1979. *The human mystery*. Berlin: Springer International.

Eccles, Sir John. 1986. *Facing reality*. Berlin: Springer.

Edelstein, Ludwig. 1967. *Ancient medicine*. Baltimore: The Johns Hopkins University Press.

Englehardt Jr., H. Tristram. 1986. *The foundations of bioethics*. New York: Oxford University Press.

Farber, S.M., and R.H.L. Wilson (eds.). 1961. *Control of the mind*, vol. 2. New York: McGraw-Hill Book Co.

Fleishman, Alan R. 1981. Teaching medical ethics in a pediatric training program. *Pediatric Annals* 10: 51–53.

Frank, Arthur W. 1991. *At the will of the body: Reflections on illness*. Boston: Houghton Mifflin.

Frank, Arthur W. 1995. *The wounded storyteller: Body, illness, and ethics*. Chicago: University of Chicago Press.

Frank, Arthur. 2001. Experiencing illness through storytelling. In *Handbook of phenomenology and medicine*, ed. S. Kay Toombs. Dordrecht/Boston: Kluwer Academic Publishers.

Gorovitz, Samuel, et al. (eds.). 1976. *Moral problems in medicine*. Englewood Cliffs: Prentice-Hall, Inc.

Hardy, R.C. 1978. *Sick: How people feel about being sick and what they think of those who care for them*. Chicago: Teach'Em, Inc.

Hare, Richard. 1977. Medical ethics: Can the moral philosopher help? In *Philosophical medical ethics: Its nature and significance*, ed. S.F. Spicker and H.T. Engelhardt Jr.. Boston/Dordrecht: D. Reidel Publishing Co.

Henry, Stephen G., Richard M. Zaner, and Robert S. Dittus. 2007. Viewpoint: Moving beyond evidence-based medicine. *Academic Medicine* 82(3, March): 292–297.

Howard-Jones, Norman. 1982. Human experimentation. In historical and ethical perspectives. *Social Science and Medicine* 16: 1429–1448.

Husserl, Edmund. 1960. *Cartesian meditations*. The Hague: Martinus Nijhoff.

Jonas, Hans. 1966. The practical uses of theory. In *The phenomenon of life: Toward a philosophical biology*, ed. Hans Jonas. Chicago: University of Chicago Press.

Jonas, Hans. 1984. *The imperative of responsibility: In search of an ethics for the technological age*. Chicago: University of Chicago Press.

Kass, Leon R. 1977. The wisdom of repugnance. *New Republic*, June 2.

Katz, Jay. 1984. *The silent world of doctor and patient*. New York: The Free Press.

Kleinman, Arthur. 1988. *The illness experience*. New York: Basic Books.

Liddle, Grant. 1967. The mores of clinical investigation. *Journal of Clinical Investigation* 46: 1028–1030.

MacIntyre, Alasdair. 1981. *After virtue*. Notre Dame: Notre Dame University Press.

Martin, Samuel P. 1972. The new healer. In *Proceedings of the 1st session, institute of human values in medicine*. Philadelphia: Society for Health and Human Values.

Pellegrino, Edmund D. 1979a. The anatomy of clinical judgment: Some notes on right reason and right action. In *Clinical judgment: A critical appraisal*, ed. H.T. Engelhardt Jr., S.F. Spicker, and B. Towers, 169–194. Dordrecht: D. Reidel Publishing Co.

Pellegrino, Edmund D. 1979b. *Humanism and the physician*. Knoxville: University of Tennessee Press.

Pellegrino, Edmund D. 1982. Being ill and being healed: Some reflections on the grounding of medical morality. In *The humanity of the ill: Phenomenological perspectives*, ed. V. Kestenbaum. Knoxville: University of Tennessee Press.

Pellegrino, Edmund D. 1983. The healing relationship: The architectonics of clinical medicine. In *The clinical encounter: The moral fabric of the physician-patient relationship*, ed. E.A. Shelp. Boston/Dordrecht: D. Reidel Publishing Co.

Pellegrino, Edmund D., and T.K. McElhiney. 1981. *Teaching ethics, the humanities, and human values in medical schools: A ten-year overview*. Washington, DC: Institute on Human Values in Medicine, Society for Health and Human Values.

Pellegrino, Edmund D., and David C. Thomasma. 1987. *A philosophical basis of medical practice: Toward a philosophy and ethic of the healing professions*. New York: Oxford University Press.

Penfield, Wilder. 1975. *The mystery of the mind*. Princeton: Princeton University Press.

Shaffer, Jerome. 1975. Round-table discussion. In *Evaluation and explanation in the biomedical sciences*, ed. H.T. Engelhardt Jr. and S.F. Spicker, 215–219. Boston/Dordrecht: D. Reidel Publishing Co.

Siegler, Eric. 1979. Clinical ethics and clinical medicine. *Archives of Internal Medicine* 139(1): 914–915.

Taylor, G. Rattray. 1968. *The biological time bomb*. New York: World Publishing Co.

Toulmin, Stephen. 1982. How medicine saved the life of ethics. *Perspectives in Biology and Medicine* 25: 736–750.

Veatch, Robert M. 1981. *A theory of medical ethics*. New York: Basic Books, Inc.

Zaner, R.M. (1988/2002). Ethics and the Clinical Encounter. Lima, OH: AcademicRenewal Press. (Reprinted from Englewood Cliffs, NJ: Prentice-Hall, Inc.

Zaner, R.M. 1992. Parted bodies, departed souls: The body in ancient medicine and anatomy. In *The body in medical thought and practice*, ed. D. Leder. Dordrecht/Boston: Kluwer Academic Publishers.

Zaner, R.M. 1994a. *Troubled voices: Stories of ethics and illness*. Cleveland: Pilgrim Press.

Zaner, R.M. 1994b. Experience and moral life: A phenomenological approach to bioethics. In *A matter of principles? Ferment in U.S. bioethics*, The Park Ridge Center for the study of health, faith, and ethics, ed. E.R. DuBose, R. Hamel, and L.J. O'Connell, 211–239. Valley Forge: Trinity Press International.

Zaner, R.M. 1994c. Phenomenology and the clinical event. In *Phenomenology of the clinical disciplines*, ed. Mano Daniel and Lester Embree, 39–66. Dordrecht: Kluwer Academic Publishers.

Zaner, R.M. 1996. Listening or telling? Thoughts on responsibility in clinical ethics consultation. *Theoretical Medicine* 17(3, September): 255–277.

Zaner, R.M. 1997. Medicine. In *Encyclopedia of phenomenology*, ed. Lester E. Embree et al. Dordrecht: Kluwer Academic Publishers.

Zaner, R.M. 2003. Finessing nature. In *Genetic prospects: Essays on biotechnology, ethics and public policy*, ed. Verna Gehring. New York: Rowman, Littlefield Publishing Inc.

Zaner, R.M. 2005. Naissances programmées? La génétique, l'aide à la procreation et le hazard d'être soi. *Esprit* (no° 2005, Décembre): 127–141.

Zaner, R.M. 2006. Benefit and mischief: Toward a phenomenology of medicine. In *A Felicidade na Fenomenologia da Vida: Colóquio Internacional Michel Henry*, ed. Florinda Martins and Adelino Cardoso, 71–84. Lisboa: Centro de Filosofia da Unifversidade de Lisboa.

Zaner, R.M. 2013. Themes and schemes in the development of biomedical ethics. In *The development of bioethics in the United States*, ed. Jeremy R. Garrett, Fabrice Jotterand, and D. Christopher Ralston, 223–240. Heidelberg/London/New York: Springer.

Chapter 3
At the Beginning and End of Life: A Meditation on the Subtle Hoax of Matter

During the early stages of medical ethics, it had become apparent that physicians were not only "supposed to pronounce 'death'," but in view of the landmark 1973 U. S. Supreme Court decision about abortion,[1] also "to pronounce 'life'"—or as the physician, André Hellegers, said at the time, the physician should "be capable of doing so" (Hellegers 1973, p. 11). What troubled Dr. Hellegers and others was that 'life' was not usually understood in public and legal discussions in the way doctors did, but in largely "unscientific" ways: as 'personal' and not 'biological' life.

3.1 In the Early Days

Medicine and medical ethics have come a long way since then. What has not changed, however, is the idea that the special concerns of medicine still must be rigorously distinguished from 'unscientific' affairs, again citing Dr. Hellegers, such as "personhood, value, dignity, or [other] words denoting societal attitudes toward biological life" (Hellegers 1973, p. 11). I say this while recognizing that most medical schools in the U. S. no longer agree with him, but have for decades offered at least something in ethics and 'values;' that hospitals, nursing homes, and other health care institutions must now have 'ethics committees;' and that some of these institutions even provide for what is called 'clinical ethics consultation.' Biomedical ethics has been a veritable growth industry since its modest beginnings in the middle 1960s—now more than ever, Arthur Caplan stated several decades later.[2] Some progress has occurred during this period, for instance, as regards certain ethical and legal issues at the end of life—formal recognition of the right of self-determination, withdrawal and withholding of life supports, 'do-not-resuscitate'

---

[1] *Roe v. Wade* 410 U.S. 113, 93 S.CT. 705 (1973).

[2] Arthur Caplan interview in *New York Times,* December 15, 1996.

© Springer International Publishing Switzerland 2015
R.M. Zaner, *A Critical Examination of Ethics in Health Care and Biomedical Research*, International Library of Ethics, Law, and the New Medicine 60, DOI 10.1007/978-3-319-18332-9_3

orders, advance directives such as living wills, and even, in some states, 'assisted suicide'—as distinct from 'euthanasia.'

Even so, on some issues there seems to have been somewhat less progress. A glance at current discussions about what to do when medical treatments are 'futile' is sufficient to make the point. Despite some clarity achieved concerning permissible withdrawal of life supports, even these discussions often remain caught in intractable ambiguity. What should be done in a situation where the physician regards further treatment as 'futile,' but the patient or significant others insist that 'everything must be done' and all treatments continued? Should the physician override the patient's or family's wishes, in effect limiting or even voiding their 'autonomous choice' to continue treatment? How, after all, should 'futility' be understood—as 'physiological' or 'personal'? Whose goals, the patient's, the family's, or the doctor's, should govern? And who should define 'benefit'—the doctor? the patient? the family? the court?

So far as issues at the beginning of life involve more than medical and scientific facts, what role do the individual's ethical (or religious, at times even legal) beliefs play in abortion, embryo research, fetal tissue transplantation, stem cell or genetic research? When we ask, 'When does life begin?' which meaning of 'life' is or should be operative? Which meaning is most important, and for whom and why? Now profoundly shaped by developments in the field of genetics—whose impact on clinical practice is perhaps its greatest hope (and surely its principal public justification)—one finds the same anomalies and disturbing questions and discussions about the human body and human person still going on: even if an embryo and a fetus are human, what is their moral status and what is the best way to define that? Which actions follow from which set of beliefs? Are things different at different stages of life? What governs (and ought to govern) the decision, for instance, about experimentation on human fetuses or embryos? What is supposed to be done in the (very common) event of serious disagreement?

These questions are not completely novel. Centuries ago, for example, Galen, the physician who more than any in Western medicine's long history left his own decisive mark on the discipline, had urged (and lamented) as he struggled (and, he thought, failed) to comprehend Plato's idea of "*psyche.*" The only viewpoint that made sense to him was the *medical materialism* long advocated by those in the ancient Greek "Rationalist" or "Dogmatic" traditions: 'soul' is a 'temperament of the body' (Edelstein 1967, pp. 173–397). If one wishes to influence the soul, one can do so only through the body. The idea that considerations other than the corporeal body lay outside the sphere of medicine is both ancient and enduring.

Although widely supported even today, the dominant medical materialism view is nevertheless peculiar. Many doctors still try to avoid confusing 'biological' notions with 'unscientific' ones such as personhood, dignity, soul, and value, for these ideas are still thought to express mainly personal, sometimes social, attitudes about biological life. Yet, doctors cannot act without being with, talking and listening to, taking care of, and helping individual patients: human beings who are *persons* in whatever sense one wishes. And how could it be otherwise, if a doctor really does wish to help not simply a body, but an *embodied person* become better, get well, or at the very least feel more comfortable?

3.2 An Overview

It is of course true that many doctors do not concern themselves with these matters, but simply go about their clinical practice without thinking about such things. Thus, despite the emergence of managed-care institutions and their already notorious intrusions into clinical situations, not to mention the always-increasing presence of governmental and insurance regulations and rules, one frequently hears about the 'doctor-patient relationship' as if it were still not much different from what it was in an early, mythic day of personal visits.

These relationships remain intimate even while largely between strangers, to be sure, and the idea that the heart of medicine is this one-on-one relation is still an article of faith among physicians, usually along with something about the Hippocratic tradition: acting on behalf of patients' best interests and doing no harm. Not without reason: not only in early times when the abilities of doctors to heal or help was highly limited, but also in contemporary societies where technologically sophisticated drugs, surgeries, and regimens are readily available, much of a doctor's daily work still occurs in such intimate relationships with patients—even if that relationship is often very brief and temporary. What Edmund Pellegrino calls the "clinical event" remains the organizing principle of medicine, (Pellegrino 1983) but has become increasingly difficult to sustain in the face of burgeoning technology and the bureaucratic structures that frame practice and research.

These peculiarities of modern medicine are probably due in large part to the fact that, since the late nineteenth century—even more since World War II—as the widely acknowledged basis for training in medicine became empirical science, especially the biomedical sciences. Despite that, however, the clinical practitioner's concrete work is still for the most part with individual patients. Eric Cassell, for example, points to the gradual but clear shift in medicine in recent times, from an enterprise focused mainly on scientific discovery to one where intervention into some patient's life is the key moment. Cassell's point is well-taken: the clinical event is still vaunted as the core point of the medical enterprise, as often by those whose practice is as much bench science as it is clinical care of patients.

It is true as well that current economic and related considerations in the United States and elsewhere tend to support the point with an increased emphasis on family practice, nurse practitioners, and the like. Despite that, however, the very success of scientifically based, technology-driven and bureaucratically organized health care makes available to even the most modest practitioner a truly awesome array of diagnostic, therapeutic and prognostic techniques and abilities—a genuine cornucopia revealing a technical complexity whose mastery and use on and for patients requires a high level of scientific knowledge was well as expert technical skills.

Inevitably, this also brought about basic changes in the nature of the clinical encounter.

It was not for nothing that over the same period of time there has been a great deal of criticism that, for all its newly won abilities to counteract the ravages of so many hitherto untreatable diseases and injuries, medicine is said to have lost sight of the 'patient as person,' in Paul Ramsey's well-known phrase (Ramsey 1970). Instead, many critics charge, it focused on disease processes, pathological lesions, biologic processes, and organ systems—more recently, genetic underpinnings, causes, statistical links and tendencies toward disease conditions. Coupled with that is the impact on medical practice and health care generally of a bureaucratic organization that encourages as much attention and time to institutional continuance as to the care of patients.

Since Max Weber's seminal writings, (Weber 1991) it has been recognized that the social, institutional organization of human social activities affects not only patients but also professional practitioners—physicians, nurses, the wide array of technicians, the institution itself and the broader social nexus—and that these organized bureaucracies tend, as has been widely observed, to 'dehumanize' and alienate, and thereby erode the governing purpose of the physician-patient relation (Pellegrino 1979). Indeed, the psychologist, Peter Lenrow, emphasized decades ago that the bureaucratic organization of human efforts to provide help to those who need help (and cannot help themselves) incorporates a deep value-conflict: while the institution is guided by mainly utilitarian values (such as efficiency and productivity), helping patients is guided by such radically different values (such as nurturing and caring)—and inevitably, it seems, the grounds for dehumanization and alienation are there as well (Lenrow 1982).

These considerations should be taken along another prominent facet of social life in those societies that provide scientific and technology-driven medicine to individuals. For the most part, doctors and patients are and tend to remain relative strangers to one another—while, however, they are, ironically and somewhat paradoxically, within a relationship that is intensely intimate and oftentimes involves profound interventions into the lives of sick and injured people. In a word, ethically speaking, the centerpiece of that relationship—reliance of the patient on the doctor's knowledge and actions—presupposes the greatest form of trust but is at the same time precisely what is most compromised and threatened by the socio-economic organization of modern medicine: situations that require a maximum of trust too often promote little grounds for trust (Zaner 1991).

Indeed, while in a sense the cherished doctor-patient relation does continue in this new world of health care, it seems to persist increasingly as a kind of vestige, more like an archeological relic than the daily reality often touted. The very words—'doctor,' 'patient' and 'relationship'—may then seem merely hollow shells, more nostalgic evocation than accurate description of what patients most need and physicians most want to provide.

3.3 Social Structure of Clinical Encounters

This is only a first step in an effort to understand what the clinical encounter is all about; more needs to be said especially about the pressing issues occurring at the beginning and at the end of life—two of the most significant clusters of issues that are posed in the 'new medicine.' In a scientifically based, technologically sophisticated enterprise designed to cure, heal or at least ameliorate pathological conditions and processes—but driven by the urge to know (research) and socially organized in bureaucratic forms—there is not much space or place for the 'patient as person.' To talk about the 'patient as person,' in other words, is a plea for more 'humanity' in medicine, an appeal already recognized in 1927 by Francis Peabody (1927) and made into a veritable icon much later by Ramsey.

A second step should at least be suggested here. Consider a typical encounter between physician and patient in a hospital. To make things manageable, we can ignore here many of its features—the specific ailment, whether the patient is self-admitted or transported by ambulance, whether the patient has a family, friends, and so on. The situation is still considerably complex. To explain a particular procedure to a patient involves specific people, each with what Alfred Schutz termed his or her own "autobiographical situation," (Schutz and Luckmann 1973, 1989) each of whom is actively engaged in an ongoing conversation at a specific time and in specific circumstances. The encounter occurs within a particular hospital and within only one of its units (emergency room, intensive care, cardiac unit, surgical ward) that includes other providers (nurses, consultants, residents, technicians, clerks), and which is only one of many units in the hospital. That hospital, moreover, is itself only one of several, sometimes many hospitals in that region, in a particular state, and in a particular country. Each of these units and hospitals operates under certain written and unwritten guidelines, protocols, regulations, and laws, the totality of which lies within the broader society with its own characteristic patterns of prevailing values (about, among other things, doctors, hospitals, sickness and health) (Fig. 3.1).

Each of the providers has his or her respective personal biographical situation, including values, beliefs, habits, and so on, and each works within a specialty or sub-specialty that has its accepted norms, codes and practices. Each practices within a specific hospital unit which has its own rules and protocols regarding resuscitation, accepted therapeutic regimens, written and unwritten rules and codes of conduct, and so on. As a socially legitimated institution, the hospital has its various rules, committees, and policies; each particular hospital region and state has its body of regulations, licensure policies, and laws; the national government has its regulations, policies, and laws; and the medical profession and specialty and subspecialty organizations have their accepted standards of admission, licensure and practice—all of which are components of the current culture with its complex folkways, mores, laws, institutions, and history. There are thus personal, professional, institutional, and prevailing social value-contexts that impact and configure each medical encounter.

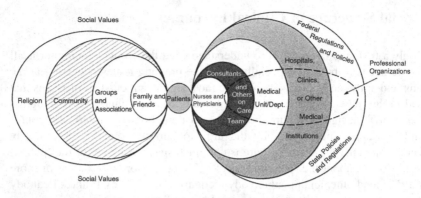

Fig. 3.1   (Zaner 1988/2002, p. 35)

There is, of course, a different but complementary complexity on the side of the patient. None of us are atoms in a void but are unique, embodied individuals with our own individual histories and biographies as well as multiple relationships, beliefs, values, and styles of life all involving other people—actions, language, situations—in a wide variety of ways. What is clear from even this brief overview, however, is that the usual way of understanding the clinical event—one person relating to another—has never been wholly true and, in any case, is not at all accurate in the context of any contemporary society.

Furthermore, with the most recent developments in the realm of medicine—specifically, the remarkable announcement in June, 2000,[3] that for all intents and purposes, the fabulous de-coding of the human genome sponsored by the Human Genome Project is now all but completed—the genome identified, its strands sequenced, and increasingly understood—the clinical event is itself becoming fundamentally transformed, even as the face of medicine's new promissory note about our genetic makeup and history is a coordinate transformation.

3.4   On Embodiment: Preliminary

Such terms as these, increasingly common though they surely are, nonetheless betray a common thematic undercurrent, a structural feature of the clinical event. However, even while this event endures as the core of what doctors' work is all about, it is more and more in danger of becoming ever more obscure and hard to define, concealing the persistent core of the doctor-patient relationship (Zaner 2005). The point deserves greater attention.

---

[3] Announced the week of June 26, 2000, both by the National Institutes of Health and the Celera Corporation—competitors even while the announcement was made jointly.

There are obviously different types of clinical encounter between helper and person needing help—whether for prevention, routine check-up, acute-care or chronic issues. Such differences are variations on an underlying thematic core. In the most straightforward terms, everyone caught up in the relationship is an embodied individual. Each person is embodied by an animate organism that is subject at once to the worldly and biological conditions affecting any such organism—illness and injury are the most obvious—and (within certain limits) to the complex strivings, efforts and wishes of the embodied person, soul or self whose body it is, who 'has' and 'lives in' that embodying body. There is nothing mysterious here; it takes no treasure-hunt to seek and find what is at issue here, for the words express precisely what each of us experiences and knows in the merest move of a finger, twitch of an eyebrow, or growl of hunger, whether any of these is specifically willed or not.

This phenomenon of embodiment is even more awesomely complex—surely as much as, in some respects even more than, the wonderful organic and genetic complexity exhibited by the human body touted by the different biological sciences. 'As much as:' after all, whatever else 'it' is, my body embodies 'me,' and in some perhaps obscure ways (Zaner 1964, 1992, 2006) is the basis for, and is itself part of, what each of us experiences at every moment of life. 'More than:' for even appreciating the marvelous intricacies of human anatomy, neurology, or physiology, there is the additional fact that must never be ignored or forgotten—that, in whatever sense may be given to the words, this one organism is experienced by me as 'mine' (Zaner 1984). It is experienced by the embodied person as uniquely singled out from among the vast universe of objects as 'uniquely belonging to me,' as uniquely that whereby 'I' enact 'my' wishes, aims, actions, and the like. If therefore we truly wished to discover the actual ground for such highly touted 'ethical principles' as autonomy, privacy, beneficence, non-malificence, respect, and others, then surely one must begin with this remarkable phenomenon—at the very least, this peculiarity of this animate organism, that it is 'mine,' cannot be ignored.

At this point, I want merely to note several of its facets. First, my body is, like any other biological entity, subject to certain of the conditions of a biological nature. As significant as that, second, is the fact that wherever 'my body' is, there am I; whatever is done to my body is done to me. In simplest terms, not only is my body susceptible to literal changes arising from the world, but it couldn't be otherwise. To say alive, I must tend to my own body: to eat, breathe, drink, move about, at once affecting and being affected by things in the world, and thus each of us always lives within profoundly intimate and ongoing interrelationships with the surrounding world and its myriad constituents—thanks to 'my' embodying organism. What can happen to objects and organisms in the world can and does happen to my embodying body. Injury, illness, handicap, compromise—all are part of what embodiment means.

Third, to be injured or become ill is to undergo personal disruption, whether trivial or more serious: if I break a leg, I can no longer 'do' as I've done before but must now 'do' differently; if I become ill, I must 'do' differently than before. If anything is truly universal about human life, then surely these must qualify: that each of us becomes ill or injured, at some point and in some manner, no matter how apparently trivial, and that we each experience ourselves as ill or injured. Moreover, to be sick or injured means that the affected person must do something about it: ignore it (but only after noting it!) and let 'nature take its course,' treat myself if I can, or turn to another for help. Whether primitive or urban, in ancient times or today, old or young, male or female, we each become ill or injured in some way, and in response very often seek the help of others—most often, someone who at least professes the ability and means to help. We become ill or injured, and we seek help to be restored, to whatever extent possible.

Drawing the sphere of attention more narrowly now—ignoring, for instance, whether the one who professes to help really can—we must note that even the simplest appeal for help is quite complex. To appeal, even if motivated by serious illness, does not mean that the sick person will accept the help offered by the other (friend, acquaintance, shaman, or physician). But *if* accepted, the appeal is marked by an initial, singular *act of trust*. And this act, it should be emphasized, even while it may be thought to require some basis, some reason for giving one's trust, yet is often enacted without that basis or reason, or with only the slightest one—perhaps, for instance, only the other's words, or the word of a friend, or some chance noticing of an ad stating someone's claim to help. If time permitted, this remarkable act should be carefully explicated, for too often nowadays not only are doctor and patient strangers, but their relationship is thought to be primarily fiduciary, with the emphasis placed on the pledge or vow of the physician (one professing the ability to help), solely on his or her trustworthiness. Clearly, however, trust must be *earned*, demonstrated or confirmed, and the initiating, positional act by the sick person is too often simply presupposed, passed over. The sick person turns to the physician first, and only later learns whether he or she is trustworthy; the fiduciary relation depends on that initial act of trust.

In any event, the relationship is one that is strikingly similar to a dialogue: an appeal by one who needs but does not know what to do, a response from one who professes to know and to be able to help, and the consequent opening-up of an ongoing relational interaction including discourse wherein the initial problem is progressively explored, continuously checked and tested, and, hopefully if at all, brought to some resolution. There are also genuine differences between these relations: the doctor's attention is understandably focused by the illness or injury that initiates the relationship in the first place, and what 'dialogue' there is may often seem merely between the doctor(s) and the patient's body, while the patient is passed over. Hence, it seems to me clear that the temptation for the physician to pay attention more to the body than the embodied person—more to the body as a biological entity than to the body as the patient's embodiment, and thus more to the disease than to the one who is ill—is not due simply to prevailing medical training or ways of organizing the doctor's work. It is, rather, a structural feature of the relationship itself: appeals by the sick, responses by the doctor—even though it is of course true that the training and institutional organization of health care enhances that temptation.

3.5 Asymmetry of the Physician-Patient Relationship

What has thus far been said is in any case sufficient to bring out what I take to be the key structural feature of the relationship: that it is, in essence, a structural *asymmetry of power* in favor of the physician, to paraphrase Lenrow (1982, p. 48). More specifically, the doctor, not the patient, has the *knowledge* and *skills* to help; the doctor, not the patient, has access to the vast array of available medical, diagnostic and therapeutic *resources* (consultants, equipment, medicines, drugs, procedures, etc.) to respond to the appeal. These are, moreover, typically backed up by considerable social *legitimation* (we are socially encouraged to take the doctor's word for what's wrong and what should be done about it), and legal *authorization* (only certain people are permitted to practice medicine).

The physician professes to possess exactly what the patient lacks: the knowledge, skills, access to resources, legitimation and authority to heal, cure, or at least ameliorate, in whatever ways. The patient is thus *uniquely and multiply disadvantaged*: by the presenting condition itself in the first place, but also by appealing to another for help and by the asymmetrical relationship itself—even though that relation must be initiated by that very person, or on his or her behalf by others (as in the case of infants, children, the incompetent, the severely ill or injured, or the unconscious). Thus, at the root of the clinical event is a fundamental set of ironies, if not dilemmas.

This imbalance toward the physician is apparent in contemporary societies in the ways health care is promoted and organized, as by the relative social value these societies place on scientific research, technological devices, and formal training of health professionals. As contrasted to the power of the physician, there is what Pellegrino calls "the peculiarly vulnerable existential state" of the patient, (Pellegrino 1982) or what I've termed the multiple disadvantages inherent to the patient's place in the asymmetrical relation. This vulnerability, moreover, is quite unique, as Pellegrino underscores:

> The poor, the imprisoned, the lonely, and the rejected are also deprived of the full expression of their humanity, so much so, that men [sic] in these conditions may long for death to liberate them. But none save saints seeks illness as the road to liberation. In no other deprivation is the dissolution of the person so intimate that it impair the capacity to deal with all other deprivations. The poor man can still hope for a change of fortune, the prisoner for a reprieve, the lonely for a friend. But the ill person remains impaired even when freed of these other constraints on the free exercise of his humanity. (Pellegrino 1982, p. 159)

This asymmetry of power and vulnerability is vital for understanding medicine, especially in its modern-day forms. Properly understood, it is also central for apprehending the at times avidly pursued political maneuverings among different medical traditions during the nineteenth and early twentieth centuries, a rivalry that eventually resulted in the endorsement of allopathic medicine and the demise or decline and continuing struggle of so-called 'alternative' medicine—osteopathy, chiropractic, homeopathy, for instance. In different terms, while the structural asymmetry pertains to any form of medicine, it obviously takes on different emphases depending on the particular medical model.

This was already understood within the Hippocratic tradition. The essential feature of the initial act of trust by the patient forms one of its central themes: "What about the patient who is putting himself and 'his all' into the hands of the physician?" (Edelstein 1967, p. 329). The patient then no less than today is "in the hands" of the healer, whether chiropractor, osteopath or homeopath, surgeon or allopathic physician. To be sure, the specific possibilities of intervention, social acceptance, and legal authorization will obviously differ, at times considerably, for each of them.

Nor were matters any less complicated in Hippocratic times, for as is not perhaps as well known as it should be, quite different traditions and understandings of the nature of medicine, therapies, patients, symptoms, and the like vigorously vied with one another for centuries. Dogmatic, Empiricist, or Skeptic, each embodied the asymmetry of power and vulnerability in distinct ways (Zaner 1992).

In any case, the widespread acceptance of the allopathic model—with its prominent emphasis on crisis-oriented, acute-care, basic science-based, research dominated, and bureaucratically organized health care—provides the context for articulating the physician-patient relationship at the beginning or the end of life.

3.6 Historical Ironies

There are two oddities at the heart of contemporary medicine that should be noted. First, we really ought to wonder why it is that what's come to be known as 'medical-' or 'bio-ethics' appeared on the historical scene when, as, and where it did. Why is it that philosophers and theologians in the United States first of all became so preoccupied with ethical issues in medicine and health care, and only certain specific issues at that, during the late 1950s and early 1960s? While some did in the past periodically express interest in medicine and its focus on sick, injured, or otherwise compromised people, it is a remarkable fact of our history that these expressions have been exceedingly rare. Today, on the contrary, we witness a veritable flood of concerns.

Indeed, since I became immersed in this field in 1970, bio-medical ethics has been a veritable growth industry—from a mere handful of people in the 1960s to standard fare on the menus of hospitals, agencies, organizations and universities (Pellegrino and McElhiney 1981). What happened? Did ethical issues only recently appear, at the very time when medicine became actually capable of doing something about illness and injury? Was there no reason for such concern in earlier times? Is medicine an inherently moral enterprise? If so, as Cassell suggests, (Cassell 1973) then why was there not the degree of overt concern in other times and places as occurred in the early 1960s in this country? On the other hand, did it only become a 'moral enterprise' thanks to certain developments in our recent history? While a case can be made for the latter, my sense is that recent developments in technology, research, or the upsurge of the rights movements, while significant in many ways, do not get at the heart of the matter. For that, we must come to fuller appreciation of the asymmetry of power and vulnerability. Before making some, unfortunately brief, remarks about that, permit me first to state the other oddity.

(a) The marriage between medicine and modern empirical science has been fateful in a number of ways; for instance, it has in part resulted in key changes in the physician's education and self-understanding. Based in a specific understanding of what science is and thus what it is to know something, medicine has become fundamentally dominated by a scientific research mentality. Indeed, it is from that view of medicine that such ethical and policy issues as informed consent, ethics review boards and committees, and other trappings of concern by anyone committed to the moral values of privacy and autonomy became so prominent.[4]

(b) When an allopathic physician encounters a patient, what typically occurs is that a specific type of *problem* is faced, which implies that a kind of knowledge is needed, therefore *research*. With that focus, the patient's particular wishes, values, beliefs, and the like, if given any credit, tend to be regarded as obviously less important than, say, physiological findings, and thus are given secondary status—which is perfectly consistent with the early Greek Dogmatic approach to sickness. In an aphoristic way, the ancient disputes between the Dogmatics and the Empirics and Skeptics—at root, whether knowing (*episteme*) or therapy (*therapeia*) is definitive of what the healer is and does—has been largely settled in our times by giving the former priority. We can still detect echoes, however, of the disputes, whether in admission committees or in medical school curriculum committees, over what really is needed to prepare students to be 'good clinicians.'

(c) But that fateful implication (and the resulting inability to say, especially in theoretical terms, what makes of physician into a 'good clinician') is not what I wish to focus upon here. Underlying that is another crucial implication that should be recognized.

---

[4] As is clear from the very fact that Abraham Flexner was commissioned to do a study, the principal result of which was a set of recommendations that directly influenced subsequent medical education and practice. The report (also called Carnegie Foundation Bulletin Number Four) called on physicians to adhere strictly to the protocols of mainstream science in their teaching and research. In 1908, the CME asked the Carnegie Foundation for the Advancement of Teaching to survey American medical education, so as to promote the CME's reformist agenda and hasten the elimination of medical schools that failed to meet the CME's standards. The president of the Carnegie Foundation, Henry Pritchett, a staunch advocate of medical school reform, chose Abraham Flexner to conduct the survey. Flexner was not a physician, scientist, or a medical educator, although he held a bachelor of arts degree and operated a for-profit school in Louisville, Kentucky. To a remarkable degree, admission to, instruction in, and later the practice of medicine was profoundly shaped by Flexner's report and recommendations.

There is a singular historical irony in the way that modern medicine proudly and in some ways rightly displays its scientific knowledge of the human body as its crown jewel, its prime if not sole concern. What is ironic is that one essential feature of this human body has at the same time been deeply obscured if not obliterated by medicine's stringent adherence to a scientific world-view: that it is *mine*, that its primary and in some ways enigmatic characteristic is that it is that whereby the embodied person is at all in the world, here and now; that it is that whereby the person embodied, whatever else 'person' may signify, at all experiences things in the world, himself included.

That fascinating irony was given wonderfully apt expression by Hans Jonas some years ago in a meditation on the prospects of a philosophical biology: in classical panvitalism, he wrote,

> it was the corpse, this primal exhibition of "dead" matter, which was the limit of all understanding and therefore the first thing not to be accepted at its face-value. Today the living, feeling, striving organism has taken over this role and is being unmasked as a *ludibrium materiae*, a subtle hoax of matter. Only when a corpse is the body plainly intelligible: then it returns from its puzzling and unorthodox behavior of aliveness to the unambiguous, "familiar" state of a body within the world of bodies, whose general laws provide the canon of all comprehensibility. To approximate the laws of the organic body to this canon, i.e., to efface in *this* sense the boundaries between life and death, is the direction of modern thought on life as a physical fact. Our thinking today is under the ontological dominance of death...All modern theories of life are to be understood against this backdrop of an ontology of death, from which each single life must coax or bully its lease, only to be swallowed up by it in the end. (Jonas 1966, pp. 12, 15)

Not a live body, one animated by and embodying a person, but only a dead one seems even remotely capable of being spliced off from the 'person' and becoming the object, *absent* the embodied person, of medicine. The point is straightforward: no science of life (bio-logy), much less a medical science focused on human life, can be well-grounded or epistemologically complete if it ignores this signal characteristic of the human body, embodiment. At the very least, such disciplines must not in principle exclude or obscure this phenomenon—and just that has occurred within the taken-for-granted acceptance of a modern scientific worldview.

Not long after Jonas' insightful remark, Ramsey had occasion to note an intriguing facet of this irony:

> In the second year anatomy course, medical students clothe with "gallows humor" their encounter with the cadaver which once was a human being alive. That defense is not to be despised; nor does it necessarily indicate socialization in shallowness...Even when dealing with the remains of the long since dead, there is special tension involved...when performing investigatory medical actions involving the face, the hands, and the genitalia. This thing-in-the-world that was once a man alive we still encounter as once a communicating being, not quite as an object of research or instruction. Face and hands, yes; but why the genitalia? Those reactions must seem incongruous to a resolutely biologizing age. For a beginning of an explanation, one might take up the expression "carnal knowledge"...and behind that go to the expression "*carnal conversation*," an old, legal term for adultery, and back of both to the Biblical word "know."...Here we have an entire anthropology impacted in a word, not a squeamish euphemism. In short, in those reactions of medical students can

be discerned a sensed relic of the human being bodily experiencing and communicating, and the body itself uniquely speaking. (Ramsey 1974, p. 59)

Concerned to evoke the "felt difference between life and death," Ramsey emphasized that this difference is felt even in the case of the cadaver: although a thing-in-the-world, it is not merely that, for it is still encountered as "once a man alive," not merely an object. I must also observe that the incommensurable contrast between life and death is dramatically encountered with the newly dead: if the cadaver evokes "gallows humor," as Ramsey notes, the mangled body lying on an emergency room stretcher awakens dread and awe. Both, however, suggest the haunting presence of a once-enlivened person—gestures, attitudes, movements, words, history—which a "resolutely biologizing age" that models the human body on the cadaver too readily suppresses or passes over.

Pointing to this phenomenon and its profound irony, Jonas and Ramsey are clearly opposed to any mind/body dualism, philosophical or medical, much less any form of materialism—both of which nevertheless are very much at home in modern medicine. But what is this "body itself uniquely speaking?" Without entertaining Jonas's or Ramsey's reflections, we should at least note that both are convinced that the materialism, mechanism, and accompanying positivism which subtly infuse much of modern biomedicine render its fundamental issue and focus, *human life*, deeply problematic. More significantly, it leaves the living human embodying organism quite enigmatic. It is little wonder, it seems to me, that such issues in medical education as what 'being a good clinician' requires are at the same time so problematic yet widely discussed[5]—or why, we saw in the last Chapter, physicians such as Samuel Martin, lament not knowing what the 'new physician' was nor how to 'train' one.

3.7 Excursus on Self

Gabriel Marcel once remarked that we often witness the appearance at the social level of language what on the other hand seems to have disappeared from our actual lives: we fervently talk about what seems to have collapsed. His example was 'personality'—a keen observation, if one considers the numerous ways in which, despite efforts to articulate what we mean, it is so lightly bandied about, associated with all manner of plain trivia, not merely in the popular press and television but, as even a cursory look suggests, in much professional literature.

Each of us is a unique self, a person, yet we are unable to come to agreement on what that means and implies. Consider: is an 8-month fetus a person? Is a 2-month fetus a self? If not, when and how does this occur? After all, if 'conception' is the significant moral moment in abortion discussions, and if being a person is

---

[5] As I learned while serving for 9 years on the Medical School Admissions Committee at Vanderbilt University.

significant as the core moment for ethics, then we must surely wonder at the cogency of such discussions. Too, is an elderly comatose individual still a person, a self? If moral value is to be placed on being a person (as so many in ethics insist, calling it 'moral agent' and connecting it to 'rationality'), then surely we ought to be clear, capable of clarity, about what that really is and implies. Are we? A brief excursus is suggestive (Zaner 2012).

If we look at any group of writings, the answer is obvious: there seem as many notions about that as there are disciplines, as Max Scheler noted early on (Scheler 1921/1961)—even worse, there seem to be as many as there are individuals who write about selves and persons. Yet self or person is often evoked in daily life, art, religion, sociology, psychology, medicine, psychiatry and psychoanalysis, not to mention other disciplines: philosophy, human sciences, and literature. If not presupposed explicitly, the self lurks importantly in the background of thought about consciousness, politics, history, religion, and in the intriguing talk about the 'patient as person.' Notions of self structure our discourse about human conduct—as in references to hallucinations, self-punishing behavior, or obsessive self-reproach. It is a phenomenon centrally invoked in all talk about human life, but in especially remarkable ways in diagnostic and therapeutic contexts. Yet on the whole, references to self and person remain oddly uncritical, implicit and unexamined.

What is 'self?' Opinions on that vary with surprising frequency and tension. There are, for instance, the great affirmers of 'substance:' Augustine, Descartes, Locke, Leibniz, or contemporaries like Macmurray. There are firmly implanted deniers: the early Sophists and some skeptics, the redoubtable Hume, and contemporaries like Sartre and Gurwitsch. Some assert, so to speak, the *tout naturel*: self is 'completely natural,' of a piece with nature, wholly indistinguishable from brain-states, complex neurophysiology, or just plain matter: the early atomists, Lucretius, Hobbes, La Mettrie, identity theorists like Armstrong, or Nobel Laureate geneticists like Sir John Eccles and Sir Macfarlane Burnett. There is also an uncatalogueable array, bewitchers of the common tongue: Heraclitus with his *Logos* forever uttering dark visions, Pascal with his tender reed and its potent *"logique de la coeur;"* Kierkegaard with his pseudonyms and puzzling 'indirect discourse;' Merleau-Ponty with his *"être au monde,"* and many others.

Nor is this all, as can be witnessed with the extraordinary variety of terms used for it: self, spirit, soul, psyche, subject, subjectivity, inner man, person; mind, consciousness, mental substance, ego cogito; as well as mental, psychic, subjective, personal, human, spiritual, and conscious life; id, superego, libido, ego, monad, transcendental unity of apperception, agent, *Da-sein, pour-soi*—the list is amazing and seemingly endless. Here, as T. S. Eliot elsewhere insightfully noted,

Words strain

Crack and sometimes break, under the burden,
Under the tension, slip, slide, perish,
Decay with imprecision, will not stay in place,
Will not stay still. Shrieking voices
Scolding, mocking, or merely chattering,
Always assail them (Eliot 1943, pp. 7, 8).

Nor is this all, for if we then ponder what it means to be embodied, equally strange and perplexing questions are unavoidable: what exactly is it that is embodied, for instance in the case of the fetus, baby, an unconscious person, or an individual in a persistent vegetative state, or someone who has undergone a brain-stem stroke (Plum and Posner 1966; Bauby 1997)? We wonder, painfully if also correctly, what sorts of actions are morally appropriate and permissible when a doctor encounters a woman with a difficult pregnancy—many of which will depend on whether or in what sense the growing fetus is, or is not yet, a genuine moral individual. The same is true when a doctor has a terminally ill patient and whose family insistently demands that 'everything be done'—when anything that could be done is regarded by the doctor as medically futile.

We may desperately wish to help; we want to do *something*, and want it to be what is right, good, and just; that, after all, is the very point and reason people become doctors, and why sick people turn to them. But what is morally appropriate and permissible, and is it the same in all situations for any individual? We want answers, to be sure, but what I want to ask is why it is that such questions are asked in the first place. To focus on that, however, only serves to draw attention to the underlying, bristling questions already noted: if 'the' human body is in the first instance *mine*, in a way *is* me, this self or person, then we have no choice but to try and get clear on what and who constitutes that self or person, who is embodied by that tiny or elderly body—so as to come to some idea of what is and is not permissible. Only then, it seems to me, is there any chance at all of making sense of the unavoidable and difficult questions at the heart of pre-natal diagnosis, embryo research, genetics, in vitro fertilization, abortion, or any of the issues that preoccupy people in so many countries. Yet, precisely those questions are not only oddly obscure, but are also deeply contentious ethically and religiously, at times breaking out into physical violence.

We place greatest value on what on the other hand seems both most obscure and least understood. Indeed, even some of our finest ethics scholars get taken in, concluding with what seem at best odd pronouncements. Al Jonsen, for instance, asks: "What constitutes the separateness that makes it possible to designate 'this person' and distinguish between 'this' and 'that' person?" (Jonsen 1994, p. 283). Note his words carefully as he argues that each of us "is constituted by a body and by certain mental phenomena associated, in a still mysterious way, with that body" (Jonsen 1994, p. 283). So profoundly intermeshed is each of us with multiple others from the earliest stirrings of life, that even the most cautious reflection seems stymied. While 'I' and 'you' are clearly different, and "where the me ends and the mine begins" seems locked in mystery, (Jonsen 1994, p. 284) he yet comes to the odd conclusion that the answer to these questions lies in the individual's genetic constitution: I *am* my genome! Jonsen argues that each of us "is constituted by a body and by certain mental phenomena associated, in a still mysterious way, with that body" (Jonsen 1994, p. 283). But that is surely inadequate, for it is not merely 'a' or 'that' body, but the individual's *own animate organism* that embodies that person. It is this phenomenon, moreover, that must be accounted for—and to allege that "I am my genome" only begs the crucial questions, since it leaves 'I,' 'embodiment,' and 'belonging' not merely "mysterious," but completely ignored and as intractable as ever. Difficult and puzzling as they may be, it is just these phenomena that are the basis of the physician-patient relationship, the clinical event, indeed medicine and biomedicine.

3.8 Conflicting Values

I began with some general observations about what are widely regarded as serious ethical issues occasioned by modern medicine. They are indeed serious, but dealing with them requires that we understand what questions are most basically posed, and why they would at all crop up in our times. My suggestion is that the pressing, even oppressive, questions centered around the beginning and end of life that so mark our times stem ultimately from what Jonas insisted is the *ludibrium materiae* and Ramsey the "gallows humor" at the heart of modern medicine and biomedicine: embodiment, whose central place in human life is in no way brought to light but is instead only further obscured by attempts to accommodate the admittedly marvelous discoveries within the new genetics.

I've also suggested that dwelling on relatively recent developments, such as technological development, doesn't get to the basic phenomena. Rather, it seems to me, it is imperative to unravel the nature and implications of the structural asymmetry of the physician-patient relationship—although this by no means is all that must be done regarding the foundations of ethics within medicine. The following considerations will, hopefully, make the point clearly.

(a) As was seen earlier, to be a patient is to be disadvantaged by the very condition that brought one to the would-be helper in the first place. Impairment compromises in multiple ways: not only by its special ways of capturing and focusing the person's attention, but also by the fact that the patient cannot 'do for herself' but must rely on others. To be impaired is to experience oneself as singularly focused on one's affliction, and thereby as uniquely vulnerable, exposed to the actions and words of others who must, unavoidably, be trusted to know and have the skills to understand and then do what they propose, so as to help. The patient is thus disadvantaged by the asymmetry of the relationship, as well as by the fact that those with power on their side are often strangers—because of which the social conditions for trust are commonly not at hand even while trust is essential even at the initiation of the relationship.

From the patient's perspective, the relationship is marked by the experience of having to rely on a great many things: instruments, drugs, procedures, regimens, and most importantly, people. To experience impairment is to find oneself in situations marked by multiple forms not only of trust but of *unavoidable trust*—especially regarding people with respect to whom, being relative strangers with the knowledge and skills to engage in highly intimate contacts, trust is itself a serious and ongoing issue. Hence, discovering and ensuring that initial and initiating trust constitutes one of the fundamental ethical issues within medicine.

(b) On the other side of the asymmetrical relationship is the physician—who has the power, skills, knowledge, resources, and socio-legal authority to judge what can and should be done, and to act. Here, several ethical issues are evident.

In the first place, many physicians and traditions in medicine's long history have traditionally taken this asymmetry as the rationale for construing the relation to patients as *unilateral* and have thus asserted their role as solitary decision-makers— a view strongly enhanced by the twentieth century marriage of medicine to the biomedical sciences and the many discoveries consequent to it. It is understandable: the more consequential the relation has become, in a sense the more pressure on those who can and do act on patients has become.

The realities of clinical work, to the contrary, force the recognition that patient encounters are *reciprocal* in that patient trust and compliance are necessary. Indeed, patients may not agree with physicians, may refuse to comply with 'doctor's orders,' and may insist on making their own decisions—including the decision to treat themselves or not be treated at all. In our times—particularly in the U.S., with its emphasis on patient rights—this difference has received strong moral and legal support. The asymmetry of the relationship does not automatically imply that it is the physician who should make decisions. It has thus become imperative for physicians to develop an understanding of the relation with patients that is quite different from that expressed in the traditional 'medical model' (White 1988). In order to develop coherent, acceptable, and practical therapeutic plans, and to enable sound decisions, physicians must learn to listen to and interpret the patient's (and family's, at times even the significant others') experiences, interpretations, meanings, and values (Cassell 1985). Along with trust, there are significant ethical issues inherent to listening, talking, and interpreting; in general terms, these are connected not simply to autonomy, but more correctly to *respect* for person's autonomy.

Second, at the very least, this suggests that the patient's place is not, and in many ways has never been, simple passivity—despite the typical usage of 'patient.'[6] The ability to alter a patient's condition and life thus does not thereby signify having *power-over*—and that, too, is morally significant. In clinical encounters, the *power-to-alter* has too often been interpreted as a sort of benign *power-for* (what was often termed 'parentalism'), or at times *power-over* or *-on-behalf-of* ('paternalism'). Increasingly, however, the physician has had to understand the *power-to-alter* as *power-with*: decision-making that requires the active participation of the patient (often, the family and others in the patient's circle of intimates). Indeed, decisions

---

[6] That is, the "quality of being patient in suffering," from Old French *pacience* "patience; sufferance, permission" (12c.) and directly from Latin *patientia* "patience, endurance, submission; quality of suffering," from *patientem* (nominative *patiens*), present participle of *pati* "to suffer, endure," from PIE root *pe(i)*- "to damage, injure, hurt".

have increasingly become the responsibility of patients or their legal surrogates (within certain limits) (Ruark et al. 1988).[7] Together with trust and respect, acting in concert with patients (with their active, ongoing consent) are surely among the reasons giving force to the idea that medicine is an inherently moral enterprise: to act *on behalf of* the patient or if nothing else to *do no harm* (which many take as 'beneficence'), requires acting *with* the patient (which suggests that the usual sense of beneficence must be rethought). The physician's actions invoke *caring* and are strictly correlated with patient *trust* and *respect*.

Third, another aspect of this was strikingly evident already in ancient medicine, even in earliest form of the Hippocratic Oath (Edelstein 1967, pp. 6–10). Reflection on this covenant, along with what has become evident thus far, makes it clear that the relationship itself makes it possible (even seductively tempting[8]) for the physician *to take advantage of* the multiply disadvantaged patient (family, household). Having the power-to-alter, the physician is obviously *able* to take advantage of that power. Well understood by the ancient empiric and skeptical physicians, a significant moral cognizance undergirded the Oath's strong injunctions: *not* to take advantage, but rather to act 'on behalf of' the patient.[9]

For the physician *to take care of* the patient, thus, signifies the moral responsibility *to care for* the specific patient encountered in each clinical situation, in large part by respecting the necessary and unavoidable trust exhibited by placing himself "in the hands" of the physician. It may well have been just this that prompted those ancient physicians who developed the Oath to emphasize that the 'Art' is governed by a "peculiar blend," as Edelstein says, of the virtues of justice (*dike*) and self-restraint (*sophrosyne*)—which several Hippocratic texts take as the core of wisdom (*sophia*) (Edelstein 1967, pp. 36–37).

Medicine is an inherently moral enterprise, then, at the core of which is a striking moral insight. Physicians are in the nature of the case involved in complex moral relationships with persons who, due to impairment and to the relationship itself, are existentially vulnerable, exposed to the power of those who wield the 'art'—and who thereby are themselves morally obligated to act justly and with restraint.

---

[7] Since the mid-1980s, it has become more common to acknowledge that the physician must recognize that the patient (family, legal surrogate) is "the true source of authority," not the physician (who acts solely as "advisor" or "consultant" to the patient).

[8] As noted in the last Chapter.

[9] The Oath's injunctions are quite specific: to refrain from having sex with the patient and members of the patient's family and household, and to refrain from "spreading abroad" what is learned in the intimacy of the relationship with the patient. Although expressed in this way, there is no reason to understand the Oath as limited to just these acts, as they express the more generic responsibility never to take advantage of the disadvantaged person, family, or household.

3.9 On Embodiment

Embodiment: that "intimate union" that so clearly frustrated Descartes and so many others after him (Zaner 1981). So intimately is self bound to its embodying organism that there is the constant temptation to say, 'I *am* my body': hit my body and you hit *me*. Here is precisely the source of 'belonging,' (Hocking 1954, Marcel 1935) from which all its other meanings are derived. Yet, however intimate and profound is the relation between the self and its body, it is equally true that its body is experienced as strange and alien.

I *am* my body; but in another sense I am *not* my body—or not simply that. This otherness of the experienced body is so profound that we inevitably feel forced to quality the *am*: it is *not* identity, equality, or inclusion. It is *mine*, but this means that self is in a way distanced from its body, for otherwise there would be no sense to 'belonging,' to experiencing it (and other things) as *mine*. So close is the union that at times it seems to 'have' me as much I it. So intimate is it that self's experience of its own-body can be psychologically surprising (its happy obedience which the self notices for the first time), even shattering (its hateful refusal to obey my wishes to do something: walk, jog, or whatever). So intimate is it that self has moments in which it genuinely feels 'at home' with it. Yet, so other is it that there are times when self treats its own-body as a mere thing that is *other* (obsessively stuffing it with food or otherwise mistreating it); or as when it is encountered as 'having a life of its own' to which self must willy-nilly attend: like it or not, 'my' hair grows and must be trimmed for certain purposes, 'my' weight goes up, 'my' bowels moved, 'my' cold cured, and so on.

Embodiment is an essentially *expressive* phenomenon. It is that whereby my feelings, desires, strivings, and so on, are enacted (albeit in culturally and histori- cally different manners). As such, embodiment is *valorized*: after all, what happens to *it* happens to *self*. As that whereby the person rules and governs, (Husserl 1961) self is at the same time subject to its conditions. What happens to it thus *matters* to the self whose body it is: the embodying organism lies at the root of the moral sense of the inviolability of self, of personhood, therefore of the 'autonomy,' 'privacy,' 'consent,' 'integrity,' and 'confidentiality' that play such serious roles in bioethics and clinical medicine. Nor does the fact that people dissemble and deceive them- selves and others belie the body's expressivity, for these are themselves expressive phenomena.

This value character of the embodying organism also helps elucidate more fully why the continuing discussions of many bioethical issues—pregnancy, prenatal diagnosis, abortion, psychosurgery, withdrawal of life-supports, euthanasia—are so highly charged and deeply personal. On the other hand, the profound moral feelings evoked by certain medical practices (surgery, chemotherapy, dialysis) and much experimentation (on embryos or fetuses, or in genetics) are understandable, as they are in effect ways of intervening or intruding into that most intimate and integral of spheres: the embodied person. Self is embodied, enacts itself through that specific animate organism which is experienced as 'its own' and is thus expressive of self.

Bodily schemata, attitudes, movements, actions, and perceptual abilities are all value-modalities by which self enacts and expresses its character, personality, habits, goals—in short, by which the self is alive as such.

As alive, the embodied self is yet intimately bound to death; when born, indeed even before birth, the individual is old enough to die. At the same time, there is something else about human birth we should note. Unless nurtured by others—most obviously, parents—a baby most surely *will* die. As the biologist, Adolph Portmann long ago observed, (Portmann 1954; Plügge 1967, pp. 34–42, 57–68) unlike other animals and many primates, human beings are born too early; most of what we are must be learned. We do not come ready-equipped at birth with the repertoire of instincts and abilities necessary to make it on our own. Sociality is thus fundamental to human life. The typical opposition, "nature *versus* nurture," is never more than a half-truth. Whatever may be our specific biological endowments, *being* is *becoming* (as Gabriel Marcel said, to be human is *"être-en-route,"* "being-on-the-way"), and becoming is a matter of being enabled-to-be by and through a myriad of actions by others. Other persons are inscribed and dwell within each of us, in countless ways producing and enabling us to be whatever and whomever we are. To become a self and eventually a person requires multiple and multiply complex interrelationships with other persons who are more fully developed (such as parents), and are also continually changing through these same relationships.

The relationships are thus *reflexive*. Kierkegaard was right: to be self is to be reflexively related to self and, on that basis, to other people. So understood, however, it is still not completely grasped, for, in his words, the self is, more accurately, *"that* in the relation *that* the relation relates itself to its own self," and as such is constituted either by itself or by another (Kierkegaard 1954).[10] As the first possibility is incoherent—such a *causa sui* cannot exist, cannot bring itself into, nor maintain itself in, being—the relation that relates itself within that relation to itself could only have been constituted by another. Kierkegaard thinks that this implies a sort of "Power" that, "as it were," lets this peculiar inwardly-outward reflexive relatedness "go out of Its hand" and lets it be on its own. Lest that capital "P" mislead, however, one can in more mundane terms remove the "P"—and we are then confronted with the phenomena of *parenting and birth*: mother, so to speak, lets baby go out from its womb (when baby is biochemically ready to exit, like it or not, there it is, *worlded*). Baby is enabled-to-be whatever it may be already within the womb, where there are already the beginnings of those mutual relationships.

---

[10] "[T]here can be two forms of despair properly so called. If the human self had constituted itself, there could be a question only of one form, that of not willing to be one's own self, of willing to get rid of oneself, but there would be no question of despairingly willing to be oneself. ... [Hence] the self cannot of itself attain and remain in equilibrium and rest by itself, but only by relating itself to that Power which constituted the whole relation" (p. 147). Despair itself, as a "disrelation in the relation that relates itself to itself," accordingly, arises by virtue of the "the relation wherein the synthesis relates itself to itself" (p. 149).

3.10   Birth, Becoming and the Ecstasis of Self

Each human being is self-aware, related to itself, however minimally this may be at any particular stage. Its being/becoming is *staged*; it is temporally phased by continually becoming itself. In different terms: *within* the relation with baby, mother is reflexively related to and as herself by means of being related to baby; baby is reflexively related to and as itself by means of being related to mother; and both are related to and as each other by being self-and-other-related within the relationship itself. There are thus two self-related individuals who are related, within their relation, to one another. They are bound to and as each other in a uniquely inward-outward manner that is the literal meaning of 'ec-stasis,' which is, as I understand him, what Jonas apparently had in mind by his philosophical interpretation of the biological concept of *metabolism* as an "ontological surprise" nature springs with living things. The living form of the organism "is never the same materially and yet persists as its same self, *by* not remaining the same matter. Once it really becomes the same with the sameness of its material contents...it ceases to live; it dies..." (Jonas 1966, pp. 75–76).

By enabling the baby to *be-en-route*, through giving birth, mother is at the same time enabled to be-herself (i.e., she is mother *by means of* and *as* birthing then mothering or nurturing baby)—and conversely. Both are profoundly marked by these reflexively complex, temporally on-going, and nurturing relationships. To be self is at root to be enabled-to-be-self-aware, to whatever extent and in whatever ways it may be. If that is so, and if it is true as well that this self-aware infant cannot continue to be without the enabling, nurturing and mutual relationships with mother, then its being *as infant* is constituted through those relationships with mother and correspondingly, mother is constituted *as mother* through her multiply, mutually-relating relationships with baby. Each, in Kierkegaard's idiom, is a *"that* in the relation *that* the relation relates itself to itself": *in* and *by means of* its very relatedness to itself it *is* self-related to the other. The already developed other person is inwardly-outwardly present within the developing self already from birth (and doubtless in subtle ways from its inception). To use a simple example to show this complexity: Robert and I are friends; each of us relates to, experiences, and values the other (I like to be with him); each relates to, experiences, and values himself within the relationship to the other (I really feel good being around him); and each relates reflexively to, experiences, and values the relationship itself (we have a good friendship).

The ontological counterpart of *metabolism*, or *ec-stasis,* is a 'being oneself *by means of* and *as* being with others.' The ecstatic being is one whose inwardness and awareness of self *is* enabled by the other. To be human is thus to be a reflexive inwardness turned (by the mutual relationship with the other) reflexively outward from the outset of life. The immediate nurturing other (parent/baby) is already within the self (baby/parent) as that whereby both are at all able to be and become. Subjectivity *is* intersubjectivity (*esse* is *co-esse*).

References

Bauby, Jean-Dominique. 1997. *The diving bell and the butterfly.* New York: Knopf.

Cassell, Eric J. 1973. Making and escaping moral decisions. *Hastings Center Report* 1: 53–62.

Cassell, Eric. 1985. *Talking with patients*, vol. 2. Cambridge, MA: MIT Press.

Edelstein, Ludwig. 1967. *Ancient medicine.* Baltimore: The Johns Hopkins University Press.

Eliot, T.S. 1943. Burnt Norton. In *Four quartets.* New York: Harcourt, Brace & World, Inc.

Hellegers, André. 1973. The beginnings of personhood: Medical considerations. *The Perkins School of Theology Journal* 27(1, Fall): 11–15.

Hocking, William Earnest. 1954. Marcel and the ground issues of metaphysics. *Philosophy and Phenomenological Research* xiv(4, June): 441–445.

Husserl, Edmund. 1961. *Cartesian meditations.* The Hague: Martinus Nijhoff.

Jonas, Hans. 1966. *The phenomenon of life: Toward a philosophical biology.* New York: Delta Books, Dell Publishing Co.

Jonsen, Albert R. 1994. Genetic testing, individual rights, and the common good. In *Duties to others*, Theology and medicine, ed. Courtney S. Campbell and B. Andrew Lustig, 279–291. Dordrecht/Boston/London: Kluwer Academic Publishers.

Kierkegaard, Soren. 1954. *The sickness unto death* (with *Fear and trembling*). Princeton: Princeton University Press.

Lenrow, Peter B. 1982. The work of helping strangers. In *Things that matter: Influences on helping relationships*, ed. H. Rubenstein and M.H. Bloch, 42–57. New York: Macmillan Publishing Company.

Marcel, Gabriel. 1935. *Être et avoir.* Paris: F. Aubier.

Marcel, Gabriel. 1951. *Le Mystère de l'être*, tome 1. Paris: Éditions Montaigne.

Peabody, Francis. 1927. The care of the patient. *Journal of the American Medical Association* 252 (6, Aug 10): 819–820.

Pellegrino, Edmund D. 1979. *Humanism and the physician.* Knoxville: University of Tennessee Press.

Pellegrino, Edmund D. 1982. Being ill and being healed: Some reflections on the grounding of medical morality. In *The humanity of the ill: Phenomenological perspectives*, ed. V. Kestenbaum, 157–166. Knoxville: University of Tennessee Press.

Pellegrino, Edmund D. 1983. The healing relationship: The architectonics of clinical medicine. In *The clinical encounter: The moral fabric of the patient-physician relationship*, ed. E.E. Shelp, 153–172. Dordrecht/Boston/Lancaster: D. Reidel Publishing Co.

Pellegrino, Edmund D., and Thomas K. McElhiney. 1981. *Teacyhing ethics, the humanities, and human values in medical schools: A ten-year overview.* Washington, DC: Institute on Human Values in Medicine, Society for Health and Human Values.

Plügge, Herbert. 1967. *Der Mensch und sein Leib.* Tübingen: Max Niemeyer Verlag.

Plum, F., and J.B. Posner. 1966. *The diagnosis of stupor and coma.* Philadelphia: F.A. Davis Co.

Portmann, Adolph. 1954. Biology and the phenomenon of the spiritual. In *Spirit and nature: Papers from the Eranos yearbooks*, Bollingen series XXX, vol. I, ed. Joseph Campbell, 342–370. Princeton: Princeton University Press.

Ramsey, Paul. 1970. *The patient as person.* New Haven: Yale University Press.

Ramsey, Paul. 1974. The indignity of 'Death with Dignity'. *Hastings Center Report* 2(2, May): 47–62.

Ruark, E.J., et al. 1988. Initiating and withdrawing life support. *The New England Journal of Medicine* 318(1, January): 25–30.

Scheler, Max. 1921/1961. *Man's place in nature.* New Haven: Yale University Press.

Schutz, Alfred, and Thomas Luckmann. Vol. I: 1973, Vol. II: 1989. *Structures of the life-world.* Evanston: Northwestern University Press.

Weber, Max. 1930/1991. *The protestant ethics and the spirit of capitalism.* London/New York: Routledge and Kegan Paul.

White, Kerr L. (ed.). 1988. *The task of medicine*. Menlo Park: The Henry J. Kaiser Family Foundation.

Zaner, R.M. 1964. *The problem of embodiment*, 2nd ed., 1971. The Hague: Martinus Nijhoff.

Zaner, R.M. 1981. The other Descartes and medicine. In *Phenomenology and the understanding of human destiny*, ed. S. Skousgaard, 93–117. Washington, DC: University Press of America/ Center for Advanced Research in Phenomenology, Inc.

Zaner, R.M. 1984. The mystery of the body-qua-mine. In *The philosophy of Gabriel Marcel*, Living library of philosophy, ed. P. Schilpp and L. Hahn, 313–333. Carbondale: Open Court Publishing Company/Southern Illinois University Press.

Zaner, R.M. 1988/2002. *Ethics and the clinical encounter*. Englewood Cliffs: Prentice-Hall, Inc. Republished by Academic Renewal Press.

Zaner, R.M. 1991. The phenomenon of trust in the patient-physician relationship. In *Ethics, trust, and the professions: Philosophical and cultural aspects*, ed. E.D. Pellegrino, 45–67. Washington, DC: Georgetown University Press.

Zaner, R.M. 1992. Parted bodies, departed souls: The body in ancient medicine and anatomy. In *The body in medical thought and practice*, ed. D. Leder, 101–122. Dordrecht/Boston: Kluwer Academic Publishers.

Zaner, R.M. 2005. Visions and re-visions: Life and the accident of birth. In *Is human nature obsolete? Genetics, bioengineering, and the future of the human condition*, ed. Harold W. Baillie and Timothy K. Casey, 177–207. Boston: MIT Press.

Zaner, R.M. 2006. It's the body that matters. In *The sociology of radical commitment: Kurt H Wolff's existential turn*, ed. Gary Backhaus and George Psathas, 155–170. Boston: Roman & Littlefield, Publishes, Inc.

Zaner, R.M. 2012. *At play in the field of possibles: An essay on free-phantasy method and the foundation of self*. Bucharest: Zeta Books.

Chapter 4
Dialogue and Trust

4.1 Appeal and Response

Encountering another person who is afflicted—whether from illness, injury, or the result of some genetic or congenital disorder—one comes into or happens on the moral order. This occurs, of course, in distinct ways in different sorts of relationships: in the present context, it occurs within the clinical encounter with its characteristic asymmetric relationship of power and vulnerability. The ground of ethics, I have thus only barely suggested, is the reflexive relatedness to and with the other person—perhaps most poignantly presented when we come upon the other-as-stranger, and even more so when the stranger is ill, when in both cases a form of strangeness comes to invade the ongoing interrelationships with other people.

Illness itself is a type of strangeness. Even while we are familiar with the variety of human afflictions, to encounter someone who is sick is to find oneself ineluctably facing something both unknown and a challenge, a dare and an appeal. With the terminally ill person, the scene can be quite dramatic: Are you still there in that body? When you become unconscious, where have you gone? Who, what, are you now? What can and should I do or say? Standing bodily before a dying person whom you do not know, what can you do or say? Perhaps all that can and even should be done is simply to be there with the individuals: mutually interrelated by means of touch, feel, word, look. Each of these only apparently simple gestures is an affirmation of the vulnerable individual as *worthwhile*. You make a difference; you *matter*.

There is more to this, for the sick individual's challenge or appeal is for a response. Most basically, the appeal is to *be-with* the sick person, to be an affirming presence of his or her continued *worth* despite impairment—which is not only alienating and debilitating, but at the same time is experienced by the ill person as demoralizing. Being with the sick person is, or can be, therefore, a *remoralizing* (Kleinman 1988). As many studies have shown, patients want most of all to know

© Springer International Publishing Switzerland 2015
R.M. Zaner, *A Critical Examination of Ethics in Health Care and Biomedical Research*, International Library of Ethics, Law, and the New Medicine 60,
DOI 10.1007/978-3-319-18332-9_4

what's wrong, but equally, whether those who take care *of* them also care *for* them—do I still matter, am I still worthwhile?

Being-with others is so pervasive in our daily lives that we rarely give it a second thought. But it has another ethically significant side to it, equally taken for granted without thinking. Especially when you encounter an impaired individual, there is a genuine appeal that you sense: 'put yourself in my shoes!' or 'look at it from my point of view!' Each of us knows this well and often takes up that perspective, as even a little reflection shows. The patient especially presents this challenge, but in so doing the patient is not asking you to *be* the patient in any literal sense. Nor is she asking you to think and feel what things would be like if you were impaired like her—though such imaginative acts can be instructive. The fact is that you are not that patient, and she is not you. None of us truly knows anyway what we would do, think, or feel, were we in that patient's circumstances literally, not unless or until illness strikes us.

To 'put yourself in my shoes' is, quite simply, an appeal to do what is most natural. Each of us is what he or she is solely within the multiple, reflexive relationships with others, each of us has been enabled-to-be what and how and who we individually are only because of those relationships. The presentation of the impaired individual is for just this reason a kind of invitation to us to be precisely be with him or her just as ourselves, as what and how and who we are—and from and with that to try and think and feel what this patient faces.

Each of us is always and essentially with and by means of others, including this unique other individual who is impaired. As mutually interrelated (as in the example of friends above), the appeal is a challenge for each of us to recognize that even in the relation to a desperately sick individual, or to one with a difficult pregnancy or an injured child, there is an intimacy, an embodied knowing that can, if permitted, affirm the what and how and who of this person, and thereby yourself. This seems to me the core of that "special occasion," that "other sense" of ethics, Albert Schweitzer identified by means of which our usually dormant moral sense can be brought to the surface (Spiegelberg 1986, pp. 219–30). More on this as I proceed.

Another word for this responsiveness to the patient's perspective is *trust* (even if, as in most clinical situations, all that is possible is a kind of temporary trust). Her words and demeanor are an appeal for us to feel-with her; they invite (and thus challenge) us to *affiliate* with her—and this act, I am led to think, is at the core of the moral order, at least within medicine. In more modest terms, this affiliative feeling or felt mutuality, the dyad care-and-trust, is *compassion*. To take care *of* a patient is, if only minimally, to care *for* the patient, to affirm that he or she matters, and this invokes affiliative feeling with the patient. It may assume quite specific forms, such as *respect* with its correlative enablement of dignity or integrity, allowing and enabling the other to be precisely what and how and who she is: Mary, with all her blemishes and failings, John with his tactlessness and, yet, dignity. She is *one who matters*; his life is worthwhile. These specific forms of affiliation, moreover, take on highly individual, embodied gestures: talking and listening, touching and being touched, looking and being looked at, thereby affirming the other's humanity and worth—in turn receiving affirmation of oneself and one's own worth. The 'matter' of embodiment, if you will, is the 'stuff' of value, not the merely measurable stuff of physical extension.

To meet the other who is ill and thereby vulnerable is to encounter a challenge to embody and enact the moral order, in the most concrete ways of the flesh. It is to enter, to 'come-upon,' the moral order, to recognize and affirm the other's appeal by means of responses tuned into the specific other person *vividly*, being bodily with the other who is bodily with me, in an act that constitutes the core sense of community. *We, thou and I, matter to one another while we grow older together* (Schutz and Luckmann 1973).

4.2 The Fiduciary and the Professional

In the last Chapter, the significance of trust was broached; it is time to look into this phenomenon more directly. Most who write about trust in clinical encounters conceive it as a fiduciary relationship (Pellegrino et al. 1991). In the following, I want to take exception to that.

The concept of the fiduciary denotes a relation that is commonly assumed to be central to professional ethics. Too often, however, discussions of the relation treat the professional in abstraction from those served (clients, patients, students, etc.). Indeed, those who are served by the professional are conceived as secondary, if not actually extraneous, to the fiduciary relation. Indeed, for some authors, the assumption of the primacy of the professional is often followed by another assumption: the fiduciary is typically taken for granted as a form of paternalism for which beneficence is the governing principle. For instance, in one of his books, H. T. Engelhardt takes both assumptions for granted. The fiduciary appears simply as an adjectival qualifier: "fiduciary paternalism," that is, the "professional judgment to determine what forms of therapeutic intervention would maximize the patient's best interests" (Engelhardt 1986, p. 281).[1]

James Childress seems similarly persuaded. The fiduciary is "another basic value in the medical sphere;" (Childress 1970) it expresses the expectation that the professional's actions will show respect for the person, and thus is understood as a form of beneficent paternalism. The main question for Childress is thus "whether trust in health care professionals to act as paternalists, that is, as beneficent decision-makers on our behalf, is warranted" (Childress 1982, p. 47).

---

[1] Although Engelhardt seems willing to accept "explicit fiduciary paternalism" as having some legitimacy in certain circumstances, his view of most forms of "implicit fiduciary paternalism" is unmistakable. The argument that there is an implicit presumption of beneficent decision making by professionals is rather problematic. Equally problematic is his argument that paternalistic interventions are implicitly agreed upon as a sort of insurance against unwise or dangerous actions by patients, which is "difficult if not impossible to defend if one takes the freedom of individuals seriously" (p. 283)—which he himself assuredly does.

This position is even clearer in Ruth Faden and Tom Beauchamp's study of informed consent. Contending that traditional codes of ethics in medicine are focused on the physician's duties or virtues, they argue that "a paternalistic or authoritarian ethics easily flowed form this." The emergence of a "language of rights," however, "abruptly turned the focus in a different direction," (Faden and Beauchamp 1986, p. 94). That is, away from a fiduciary relation and toward informed consent, to discourse centered more on autonomy, entitlement, and rights than on beneficence. In any event, to think about the fiduciary is for such scholars to think about paternalism, and this concerns the professional first of all, if not exclusively.

It may be that the initial emphasis on the place of the professional, and the de-emphasis on that of those served, is key to the tendency to treat the fiduciary as a matter of paternalism. The danger in this approach is that it risks conceiving professional ethics as unilateral, whereas it is to the contrary reciprocal: clients or patients are quite as essential to the relation as are professionals. In somewhat different terms, client or patient trust is crucial, indeed fundamental, to the understanding of professional trustworthiness.

4.3   Trust and the Trustworthy

With some qualification, Robert Sokolowski's analysis seems to make the same assumptions; it thus risks muting the place of initial trust in favor of trustworthiness and the fiduciary. That the professional and client relation is fiduciary means for him that the client subordinates some limited part of himself, his prudence, to the professional. The main question thus concerns professional trustworthiness, which can most often "in principle" be assumed from the fact of his/her having been certified as a professional. Thus, Sokolowski believes that there is

> an elegant anonymity to professional trustworthiness; if I get sick away form home and must go to the emergency room of a hospital, I can in principle trust doctors and nurses I have never met before. . .because they are presented as members of the medical *profession*, persons who are certified by the profession and who can, *prima facie*, be taken as willing to abide by its norms. (In Pellegrino et al. 1991, p. 31)

Thus, the fiduciary concerns the professional first of all, specifically his or her trustworthiness, which can "in principle," be trusted precisely because of the "elegant anonymity" of socially approved certification. It thus is clear that *client trust* is a function of professional trustworthiness; it is, in a way, the guarantee for the former: a patient can trust the professional in view of the elegant anonymity of the latter's trustworthiness. For all that, however, the client remains the "ultimate agent" since "the professional assumes responsibility for only a limited part of the client's life" (In Pellegrino et al. 1991, p. 27).

It might be noted that Childress may seem an exception: at one point, for instance, he argues that "if it is effective, paternalism presupposes trust. . . ."

(Childress 1982, p. 47). It is not precisely clear, on the other hand, just how "presuppose" or "effective" are meant, and much depends on that. Nor is it at all obvious that the fiduciary relation is necessarily paternalistic; this is simply taken for granted by these authors. In any event, the more cogent analysis surely ought to focus on the *relation* itself; that there is a relation is presumed by the relata—that is, by the presence of a professional and a patient or client—hence in some sense client trust and professional trustworthiness are, minimally, both necessary to the relation and are reciprocally related. What needs clear and cautious analysis, however, is what is too often simply taken for granted, the place of the client or patient in the relation: the phenomenon of trust itself. This must not be swallowed up by the professional's place in the relation, for then there ceases to be a relationship.

4.4 Patient Trust

It is necessary to emphasize the usual way of conceiving these matters, for then it is becomes very clear what is taken for granted: the fiduciary relation from the perspective of the patient or client, that is, the *phenomenon of trust* itself. As I've pointed out elsewhere, in fact, the Hippocratic tradition in medicine places patient trust at the heart of the physician-patient relation. For just this reason, moreover, I've taken the mythic figure of Gyges as the polar opposite of the core Hippocratic discipline. Given that the patient is at a distinct disadvantage, and the physician at a distinct advantage, the one with power on his/her side is the physician, who is as such constantly haunted, even tempted, by that power. Why not take advantage of the vulnerable sick person? The Hippocratic tradition is perfectly clear: precisely because of the patient's vulnerability. For whatever reason the patient turns for help to the would-be healer—whether in light of that "elegant anonymity" that impresses Sokolowski, or something else—there simply can be no relationship, fiduciary or not, with the healer. Hence the relationship must needs have already been initiated by the patient turning to the healer, and that act of initial and initializing trust is the crucial aspect of the phenomenon that requires understanding. More on this significant theme at a later point.

I take it that it is just such considerations which lie behind Edmund Pellegrino and David Thomasma's key idea: after showing that medicine is an inherently moral discipline, they insist that an "ethics of trust" is its essential feature (Pellegrino and Thomasma 1981, p. 67). With its focus on the vulnerable, sick person, indeed, the classical axiom of medicine since Hippocratic times—"to help or at least to do no harm" (In Jones 1923, p. 165)—makes the initial trust by the patient the central requirement of the fiduciary relationship. To paraphrase Kant, the fiduciary without patient trust is empty; and, it may also turn out that trust without the fiduciary is blind: as suggested, these are strictly reciprocally interrelated.

Elsewhere, Pellegrino argues that the physician "takes upon himself" the responsibility of taking care of the sick, injured, or debilitated, "not as a negotiated task but as an imperative built into the very nature of clinical medicine" (Pellegrino 1983, p. 164). This imperative, I am suggesting, derives from the essential vulnerability of the patient, who places himself (or is placed by others) "in the hands" of the physician in a total response of initial trust, and that it is at once uninvited and sourced in the patient's vulnerability gives it all the more force and depth.

These matters are, however, more complicated. This is perhaps clearest when it is recognized that what Pellegrino apparently grants at one point he then seriously qualifies. The physician probes and even violates the patient's body in ways not permitted even to those whom the patient loves, and because of this Pellegrino understands trust as grounded on the trustworthiness of the physician, but does not go on to appreciate the fact that the patient must, as I will shortly point out, place trust in the physician without in the first place knowing whether or not the physician is indeed trustworthy—hence, trust cannot be grounded on the latter.

In their book, Pellegrino and Thomasma argue, "the axiom of care for the vulnerable individual is the ground for an ethics of trust...between doctor and patient" (Pellegrino and Thomasma 1981). The crucial phenomenon, therefore, is quite clearly the vulnerable person's act of placing himself "in the hands" of the professional: this act receives it proper emphasis only when seen as based in the "axiom of care," that is, physician trustworthiness.

I will return to this specific point later. For now, it is important to note that, in order to enter into a professional relationship with a physician or other professional providing some form of expert help to others, is to enter a domain that is already textured by multiple forms of trust on the part of the vulnerable individual. Here, a distinction of some significance is needed. In a sense, it is true enough to say that one can trust doctors, lawyers, teachers, and the like, even when they are total strangers, thanks to that "elegant anonymity" of professional trustworthiness noted by Sokolowski and captured by that signed official document regularly found on the doctor's office wall. The sheer fact of having been socially "certified" as professionals means that clients, patients and the like will typically take it for granted that professionals are "willing to abide" by the norms of the profession.

But note that this omits something quite crucial: the phrase, "I can in principle trust," can only mean that the professional's trustworthiness is *typified and thus typically taken for granted*. In other words, here, "trust" refers strictly to what is taken for granted as part of the typified knowledge each of us has just so far as we are members of the same social world and culture. Probing into specific clinical encounters, however, invariably confronts us with a quite different sense of trust. With illness, injury, handicap or other compromising condition that prompts a visit to a physician, for instance, the patient presents not only specific sorts of bodily and/or mental distress, but also personal suffering and anxiety: any disease is at the same time a "dis-ease." An essential component of that personal dimension is that, to one degree or another, the person can no longer, by the very fact of illness, take for granted much of what he or she had hitherto been taking for granted: precisely this is compromised and brought into explicit awareness by the debilitating condition. Indeed, that this occurs is a key part of the meaning of the illness experience.

To undergo illness or other form of need sufficient to bring one to a professional is to find that one does not know or cannot do for oneself, and can no longer take just that knowing and doing for granted—whether the action is mowing a lawn, writing a letter, or conversing smoothly with a friend. In fact, one of the most common themes of the fiduciary relation is that even the professional's typically taken for granted (and "elegantly anonymous") trustworthiness is itself an *issue*, a *question*, for the patient.

Accordingly, a key part of what the illness experience means to the patient is whether trust, even though in many ways unavoidable, is actually warranted; at the very least, this is a question at the outset of the relation to the professional. While it may be that this theme tends to become a more explicit question with more grievous illness—or when the sick person believes things are serious—that is not always the case. Even when the need for help seems or is in fact less serious, many people still express the question, albeit often in more subtle and muted ways.

Illness or being in need of help from the professional other, furthermore, invariably includes various types of uncertainty, which texture every individual encounter. The essential component of the uncertainties any patient experiences arises from the experience of one's own vulnerability, of having little if any choice but to trust or place oneself "in the hands" of the professional.

4.5 Forms of Unavoidable Trust

From the patient's perspective, therefore, the professional's trustworthiness is closely tied up with various forms of unavoidable trust, the very fact of which can only enhance the tensions already ingredient to illness, including no longer being able to take for granted one's typical ways of relating to other people—who, very often, are strangers and thus who unavoidably enhance those tensions. This phenomenon needs more careful explication, as it has considerable significance for the fiduciary relation.

Consider merely some of the circumstances a sick person faces. As illness variously impairs the ongoing, integral connection of body and self, so too does it alter the ordinary relationships with other people and the surrounding world of things and events. As any of these relationships are more or less disrupted, the patient finds him/herself involved in various kind of unavoidable trust.

Patients have no choice, in a sense, than to trust not only their doctors but a multitude of others as well: nurses, laboratory technicians, researchers, administrative personnel, manufacturers, transporters, and countless (mostly anonymous) others (Hardy 1978). They also have no choice but to trust a great many things: for instance, any healthcare professional must trust that the material used to repair body-parts is appropriate, as also the bandages, drugs, surgical equipment, and still others. They also have to trust in the efficacy of numerous procedures: sterilization, administering of anesthesias, surgical techniques, referrals, preparations of drugs, and so on.

Having no choice but to trust in these and many other ways, communication among those involved in any clinical encounter thereby becomes highly important. A man with lung cancer, for example, emphasized: "when the doctor told me I had this tumor, frankly, it alarmed me, but he did it in such a way that it left me with a feeling of confidence" (Hardy 1978, p. 9). A diabetic patient underscored the point: "if you can't communicate and you can't understand your disease, then you don't have confidence in the medical help you are getting" (Hardy 1978, p. 236).

4.6   Unavoidable Trust and Uncertainty

Illness of itself provokes a need to know and to understand: what's wrong? Is it serious? What does it mean to and me, for my family, now and in the future? Is my condition curable or only treatable? If treatable, by what means, at what risk, and at what cost? What should I do?

Clinical judgment includes several distinguishable (although inseparable) phases or moments. In Pellegrino's analysis, clinical judgment answers to three major questions: what is wrong? (diagnosis), what can be done about it? (therapeutic determination), and what ought to be done about it? (prudence) (Pellegrino 1979). To these, however, it is imperative to add the classical sense of prognosis, for every illness renders the future into sharp questions: what's going to happen? How long with I hurt? How will my family be affected? Will I be able to work? Am I dying?

Each of these moments involves some form of uncertainty and ambiguity—which signify necessary fallibility on the part of the professional. As Robert Hardy (a hospital administrator) discovered in his numerous interviews with patients and their loved ones—before, during and after hospitalization—most were concerned not only to know and understand their medical problems, etc., but equally to know that those who *take care of* them also *care for* them. More particularly, the sick person concretely experiences his or her body as a source of uncertainty: for instance, what is causing pain, how long it will last, what it signifies now and for the future, and so on. Precisely in view of these multiple uncertainties, there is always an initial, serious question for every patient: with so much at stake, is trust in *this* physician truly warranted?

The patient has a profound and understandable desire, thus, to know with as much clarity and certainty as possible what can be done and should be done, and whether those who take care of them also care for them: whether one can trust those who have and communicate their knowledge and concern, and eventually act on and/or to the patient. Trust by the patient is always in some way and to some degree set within such questions and thus is given only within a sense of uncertainty. If there is, as Pellegrino and Thomasma have urged, an "ethics of trust," it must, therefore, be set first within an *ethics of uncertainty*.

It is true that where there are specific and effective treatments for some diseases (for instance, penicillin for pneumococcal pneumonia), trust is often unproblematic or muted, and the typical presumptions about the professional's conduct remain for the most part unchallenged—unless, of course, something in the physician's conduct or attitude forces some presumption to surface as a question once again. Clinical judgments are often not so clear, however, especially with regard to the major diseases of our times—heart disease, cancer, stroke, diabetes, etc.—which often can only be managed, not cured. Clinical judgments are even less clear as it is increasingly imperative to reckon with and take into account the person's experiences of, and meanings given to, any illness.

As issues like these multiply and grow in uncertainty, so do the chances of compromise to the patient's unavoidable trust. For patients, it is hardly appropriate merely to cite statistical probability patterns for classes of diseases and persons. Although thought of as ways to evoke trust, such conversations often backfire. For the patient, uncertainty and ambiguity more often have the sense of being-at-a-loss, being adrift and unable to take one's bearings and to know what to hold by—which is to discover that trust is itself the central, critical issue in every clinical encounter. To know what can be counted on is to know what can be trusted, and if the one fails, so is the other compromised (Schutz 1964, p. 97).

The illness experience thus makes prominent the need for candid, sensitive conversations that can evoke and warrant, even if only temporarily, genuine trust so that decisions, at times vital and irreversible, can be made even when their basis may remain incomplete, uncertain and ambiguous.

4.7   Unavoidable Trust and Strangers

For a patient to enter the world of medicine is to enter sometimes forbidding and alien environs—such as large hospitals or clinics—even when the doctors and nurses may be familiar to the patient and/or family. Mostly, however, strangers more often than familiars surround the patient: other patients, families, visitors, hospital personnel, and still others. Patients also become surrounded by the strangeness of things, equipment, buildings, schedules, food, procedures, and an array of anonymous people working and producing those goods and services found in every health care institution.

Sociologically and even architecturally, hospitals and clinics seem designed more to enhance than to ameliorate these forms of strangeness. Stripped of familiar things (clothes, possessions, surroundings) and told they must wait while clothed in anonymous garb that provides ready access to body parts and places otherwise forbidden to other people, even loved ones, patients are then asked to discuss the most intimate details of personal and bodily life to whomever may by chance be assigned to them. Their illness narratives, histories and personal features become quickly converted into 'cases' openly discussed by hosts of anonymous others—doctors, nurses, students, therapists, aides, administrators, pastors, ethicists—with confidentiality and privacy very often little more than words on a document, merely faint reminders of private places and personal details. They are poked and prodded, swabbed and stuck, palpated and auscultated, in intimate and even humiliating ways, all in the service of being taken care of, and in whose necessity and efficacy they must simply and unavoidably trust. To be a patient, one must be patient indeed; to trust, however, surely requires more than mere patience.

Illness itself is alienating, rupturing the person's usual ways of feeling, acting, moving, and integrating body and self (Gadow 1981, p. 88).[2] Indeed, a crucial dimension of trust—trust in one's own body—is existentially breached by illness or injury, the person's bodily experiences taking on a kind of interior strangeness. New and peculiar bodily feelings emerge, for instance, and are often very difficult to at best to convey in terms understandable by doctors and nurses. Illness disrupts the usual routines of daily life, including the ways in which the person typically relates to others. Even if the person is gently encouraged to talk about these sensitive and often furtive feelings and relations—is it the disease, the drug, the hospital or me that makes doctors or nurses seem so intimidating?—patients must trust that the professionals are correct when they insist that the discourse is important, and that they will not only hear but listen, not only understand but be understanding.

When communication of this sort is among strangers, the experience can be even more confounding and tricky—which, as much as anything else, compromises patient trust, even when one knows in typical ways that these professionals may "in principle," as Sokolowski insists, be taken as willing to abide by such norms as confidentiality—a presumption that, patients and families may soon realize, is frequently open to suspicion in any complex, bureaucratic institution of health care, where one's most intimate secrets are easily accessible to many people one rarely bargained for on admission (Zaner 1994, pp. 1–11).[3]

In a society in which relationships among strangers predominate, communication tends to be designed more for temporary ease of social passage and commerce than for intimate probings and disclosures of secrets. For the hospitalized patient, matters can become acute, especially regarding trust: talking and listening are often merely exercises in remoteness with only the outer shell—the words merely—of intimacy, and thus can be more sham than real.

And thereby, it must be emphasized, the "elegant anonymity" of professional trustworthiness may swiftly fly out the nearest exit. Yet, it is precisely the sick person and family who need most of all to know what's wrong and who must, unavoidably, trust that those who take care of, nevertheless genuinely care for him or her—a major index of which is candid, continuous, and sensitive discourse about findings that, one hopes, will be kept confidential.

---

[2] Sally Gadow has pointed out that as "the felt capacity to act and the vulnerability to being acted upon," the lived body is precisely the "primary being-in-the-world that is ruptured by illness or injury." (p. 88).

[3] It was swiftly clear to me in my own clinical consultations, for instance, that few if any had ever bargained on meeting a philosopher, much less that I would have such easy access to the most intimate details of their lives.

As Alfred Schutz has vividly shown, commonsense life consists of a relatively well-organized set of typifications (of people, things, events, etc.). So long as things remain more of less unruffled, for the most part we simply take it for granted that our typical and typifying ways of thinking and acting are, generally and for the most part, assumed without question to be correct for all practical purposes. That is, we take it for granted, Schutz emphasizes (a) that "life and especially social life will continue to be the same as it has been so far," (b) that we may continue to rely on what has been handed down to us (by parents, siblings, teachers, traditions, etc.), (c) that in the ordinary course of affairs it is sufficient merely to know *about* the general style or type of events we usually encounter, and (d) that neither our typical ways of acting, interpreting, and expressing ourselves nor these underlying assumptions "are our private affair, but that they are likewise accepted and applied by our fellow-men".

The point is obvious: if any of these assumptions fails or becomes openly questionable, commonsense ways of thinking and acting also become unsettled: a crisis occurs, what hitherto 'worked' works no longer, etc., texturing the encounter with the stranger. Schutz notes that the stranger is "essentially the man who has to place in question nearly everything that seems to be unquestionable to the members of the approached group" or individual member of the group. He has not participated in their cultural life and history and thus, "seen from the point of view of the approached group, he is a man without a history" (Schutz 1982, p. I: 97). In the case of illness, therefore, the act of "history-taking" is far more than of mere medical interest. The patient may know *that* the approached group (the doctors and nurses, for instance) has its own ways and routines, but these are not an integral part of the patient's own biography. Hence, what is taken for granted by the one cannot be taken for granted by the other.

By the same token, from the point of view of those 'at home,' while the stranger is perhaps seen as having some sort of culture and a history, perhaps even a personality, these are precisely what is not known in any detail (or, it is known only in typified terms), hence those at home cannot take for granted regarding the stranger what they otherwise typically take for granted regarding one another.

When what brings the 'newcomer' (patient) to approach those 'at home' (medical professionals) is something critical like illness, moreover, things can be exceedingly difficult and baffling, and a source of contention. What is otherwise typically regarded as obvious and settled within the 'at home' group (for instance, the use of initials such as 'PTA' = 'prior to admission', as opposed to 'PDA' = 'patent ductus arteriosis,' may augur horrific implications to a patient and family) comes into question for the stranger. Thus, the familiar and routine, including common language, may then no longer be settled and regular. It is then evident that the meanings of trust and being trustworthy are significantly different from situations where people are less strange to one another.

Furthermore, in the best of times all that may be hoped for among strangers are situations in which only temporary trust is possible (Lenrow 1982). Even that, however, must be earned, since there is little basis for trust among strangers; and lack of the conditions that allow for, much less promote, trust can only mean that a vital part of the therapeutic relationship is missing or threatened.

4.8   Trust and the Professional's Power

To be a patient is thus often to find oneself in a deeply ironic predicament: actions of touching, feeling, talking, and probing, which typically attest to personal intimacies, now go on between strangers, and thus have a very different significance in the relationship between professional and patient. For the same reasons, professional trustworthiness, "elegantly anonymous" or not, is as often problematic even while it promises the very sort of expert help a patient or client seeks.

It has been noted that there are compelling difficulties both patient and doctor experience while trying to communicate and to preserve even minimal conditions for trust within the complex bureaucratic institutions of health care, although this is still only a glimpse at the fundamental asymmetry of power in favor of the professional that characterizes the relationship with patients.

Understandable and unavoidable, at the same time the asymmetry itself is a disadvantage to the patient. The mother of a partially sighted girl, for instance, emphasized how "overpowering" physicians can be: "They've got an edge on you," (Hardy 1978, p. 92) and a surgical patient wondered: "You're all in their hands, and if they don't care for you, who's going to?" (Hardy 1978, p. 40). In this respect, Pellegrino seems right on target when he stresses that patients "are condemned to a relationship of inequality with the professed healer, for the healer professes to possess precisely what the patient lacks—the knowledge and power to heal," or at least to help ease discomforts, pains, and so on (Pellegrino 1981).

As emphasized earlier, this inequality is nevertheless constitutive of the helping relationship. Doctors have special knowledge and skills (won through education and training) in the ways of the body (sometimes, the mind as well). They have access to resources (people, technologies, medications, institutions, funding) and are socially and legally legitimated and protected (licensure statues, professional memberships) to act on behalf of sick people. Indeed, it is thanks to the socialization, cultural distribution and common acceptance of the asymmetry by the members of the society, that one can at all say that 'I can in principle trust' professionals; that through being 'certified' one can take it for granted that they can be trusted to a bide by its norms, and thereby be experienced as trustworthy.

Strongly enhanced by formal institutionalization, social legitimation, and legal authorization, this inequality of power is intensified by the sick person's illness and vulnerability. The patient is a supplicant whose appeal is precisely an endorsement of the very phenomenon that constitutes the inequality: the ability to know and the skills to treat, heal, possibly even restore. All of which is rendered more problematic when the participants are strangers to one another and can take neither trust nor trustworthiness for granted—that is, cannot automatically assume that they share values, attitudes, desires, aims, and the like, in such a way that it is warranted to believe in, to trust, the professional's trustworthiness.

4.9 Responsiveness to Trust

To be a professional in the proper sense is find oneself being trusted without there being necessarily any good reason for this. It is furthermore to profess the ability to help those outside the membership of the profession (and even those inside it, in the event they come into the kind of difficulty that calls for professional intervention), and this is fundamentally to embody a promise to those in need of help and who have exhibited initial trust in the professional.

While all the usual responsibilities of making and keeping promises hold here, there are crucial differences for the fiduciary relation between professed helper and person in need of help. The professional promises to be the very best he or she is capable of being for each and every client or patient, where 'being capable' signifies not only knowledge and technical competence. He or she also promises not only to take care of, but also to care for the patient and family—enough at least to warrant trust—that is, to be candid, sensitive, attentive, and never to abandon them. What is promised in the response to initiating trust is that the trusting person will never be taken advantage of; and not only that the professional will seek to understand the presenting condition (physical ailments, physiological condition, etc.) but also to be understanding of the patient's situation and needs.

While the asymmetry of power in favor of the professional presents the constant temptation of taking advantage of the vulnerable patient—precisely because that temptation must be understood in moral terms—that very asymmetry imposes quite special obligations and responsibilities on the professional. Not only must the power be used competently and fully, but it must never be misused or abused—specifically included in the Hippocratic oath within the interesting forms of 'mischief:' there is a vow neither to do mischief oneself, nor to let patient's do mischief to themselves. The professional's abilities and knowledge are placed at the disposal of patients and their families—and herein lies, I believe, one of the central sources of what Gabriel Marcel analyzed as a core moral phenomenon: *disponibilité*: being at the disposal of those whose conditions place them asymmetrically at a disadvantage (Marcel 1940). These special responsibilities arise to a significant extent from the existential situation, the vulnerability, of the patient and family generated by both illness and the asymmetry (Liddle 1967).

The issue is all the more acute just because the sick or injured person is uninitiated in the art of medicine, does not know what is wrong nor what to do in order to find out, much less what to do about it. Moreover, the patient does not even know how to assess whether the professed healer is capable of healing—at least not until or if that has been achieved, to whatever degree. But however the patient tries to be more certain of the physician, physicians must ask themselves whether the initial trust the patient places in the doctor (knowledge, skills, ultimately the person himself) is at all warranted? Paradoxically, trust seems able to be warranted only after the fact, even while it is existentially required beforehand.

4.10   Trust and the Professional's Duty

By providing some texture to the fiduciary relation, noting in particular how the patient or client is unavoidably situated within various forms of vulnerability, I hope to have made prominent the phenomenon of *initiating trust*, as I may now call it. Patient or client trust is quite as essential to that relation as is professional trustworthiness; it is, indeed, that to which the professional must from the outset be *responsive* and for whom he or she must be *responsible*.

The point can be expressed differently. In most cases, the patient or client is the one who initiates the relationship with the professional—although there are obvious exceptions. Without the universal fact of illness, injury, or other debilitating condition, there is no call for a physician. The patient or client experiences some form of real or imagined lack of or functional failure in knowledge and ability to do for oneself, and turns to someone who professes the ability to help: the professional. Integral to that need or lack is a break in the person's otherwise taken for granted knowledge and ability, which frequently carries with it one or another degree of disruption in the person's usual ways of relating to others.

Just because of these various disruptions or failures in what is usually taken for granted the professional's *first and continuing task* is clear: responsibly attending to each individual's specific concerns, experiences, self-interpretations, and functional lacks or failures. The concept of responsibility is thus indispensable to the professions, and in the complex sense intrinsic to it: *to be responsive to* each individual within his or her unique circumstances and condition, *to be responsible for* whatever is then said or done in response to and on behalf of that individual, and *to be responsible to and for* the conduct of other professionals and, indeed, the profession itself.

The professional as such takes on the vital task of *enabling* each individual patient or client to be restored in respect of the needs that brought him or her into the professional's sphere of action and knowledge in the first place. The professional has the responsibility of helping the person, ultimately seeking to enable the individual to be restored to him- or herself (to the extent possible in each instance).

In this respect, there is a significant *methodological injunction* for the professional that harbors vital moral content: as far as possible, *take nothing for granted*, patient or client trust most of all. The professional has the moral imperative of recognizing that the typically taken for granted trustworthiness typically attaching (with "elegant anonymity") to professional status by the sheer fact of social certification is itself the first and most pressing *issue* of any fiduciary relationship.

This, as I've urged more than once, holds true regardless whether or not the fiduciary relation, as Faden and Beauchamp argue, should be conceived in terms of "valid entitlement" and "rights." They contend that things have changed in recent decades, and that "the new kid on the block in medical ethics" is entitlement. Although "usually reserved for law. . .it literally invites replacement of the beneficence model with the autonomy model" (Faden and Beauchamp 1986, pp. 94–5). Thus, whatever may have been its origins in simpler times, informed consent is a matter of rights and entitlement, and this is the core, they argue, of the fiduciary relation.

One should not for a moment concede that point, however, for even if trust and trustworthiness were shifted into entitlement and informed consent language, all the themes and issues of the first only reappear with the second—and with even greater urgency, since trust tends to be suppressed in favor of autonomous action, a crucial mistake, and a solid reason "autonomy" cannot replace trust in any sense.

The point is clear enough regarding confidentiality. But so is it with informed consent, most especially when this is, as it should be, understood as an ongoing process between persons who are in the nature of the case asymmetrically related. In order for a person to be appropriately 'informed,' much less not coerced and allowed to be free in giving 'consent,' trust remains the *sine qua non* for both the professional's disclosure of pertinent information (such as risks), as well as the actions proposed then carried out. Professional trustworthiness is still a vital *issue* in constant need of warranting trust in every individual encounter, whether or not informed consent is brought in as way to provide formal guarantees (or the threat of subsequent legal recourse).

4.11 The Phenomenon of Trust

The attempt to shift the discourse away from beneficence to entitlement and autonomy does not in the least get rid of the phenomenon of trust; such discourse simply buries trust and presupposes that its sense is already established. As I've emphasized many times, trust cannot be predicated on professional trustworthiness, for this, too, presupposes trust on the part of the patient regardless whether it is "elegantly anonymous" or not. In short, any such move simply obscures the fact that trust is then made to appear as only a part of the typified and taken for granted recipes for knowing and acting in the everyday social world—precisely what must never be taken for granted, and which illness experiences positively prohibit from taking for granted, whether these are recognized as such or not. Professional trustworthiness is therefore more accurately characterized as the indispensable *promise* that trust is an issue, that it must be warranted, and is always an issue to be won or lost solely from what transpires within the course of the clinical encounter.

Trustworthiness, the fiduciary in general, is an issue and a promise precisely insofar as it is essentially a response to someone's appeal for help; the one, professing the ability to help, responds to the other, standing in need of and asking for help. Central to an ethics for the professions, therefore, is the idea of *specific responsibility*: responsiveness-to and responsibility-for, defined within each context. To articulate this vital moral idea, the phenomenon of *initiating trust* is essential. To get at this phenomenon, and the ethics connected with it, requires that we probe into the phenomenon of appeal (and that of response) at the heart of the fiduciary relation.

Although the situation of the one in need of help doubtless varies depending upon the specific kind of need at issue, it seems clear that *existential vulnerability*, to whatever degree it may be, is a common theme. The patient or client, student or penitent, exposes him/herself existentially as being unable to do whatever it takes to correct what has gone wrong (illness, injury, etc.). In this, therefore, as Sokolowski rightly notes, "when I approach a professional, I submit something more than a possession of mine to his expertise; in a distinctive way, I subject myself and my future to his assessments and to his judgment" (Sokolowski 1991, p. 27). Handing over the "steering of my life," to whatever degree, "I let him take over not just one of my things, but my choices and activities themselves," something for which my own knowledge and resources have proven inadequate (Sokolowski 1991, p. 27).

The other side of this "handing-over" is that I am myself at stake; I am disadvantaged, I am at risk, in a most concrete way within the relationship with the professional, for my vital wherewithal and abilities are insufficient to see me through and I appeal to one who professes the ability to help in the manner and the kind of help I need (and who is 'credentialed'). An essential component of the person's vulnerability, therefore, is this exposure of self, the person's being unequipped to carry through on his/her own, thus his/her inequality before the professional.

That vulnerability, I believe, is the vital central of any clinical ethics—perhaps any ethics whatever—and is experienced by the person as a compelling appeal to and need for the other's responsiveness. This vulnerability, furthermore, has a quite particular format: the appeal for help is a call to the other, the professional, to *affiliate* with me, to feel-with and understand me from within my own appeal and specifically vulnerable state. Consider again the example of illness, now from the perspective of the physician who is called on to be responsive to the patient. Not only must he or she be able to stand back, analyze, classify, measure and reason, but also, as Pellegrino recognizes, to

> feel something of the experience of illness felt by this patient. He must literally suffer something of the patient's pain along with him, for this is what compassion literally means. Often the physician heals himself while healing the patient; oftentimes he cannot heal until he has healed himself... (Pellegrino 1983, p. 165)

But we need to be very cautious about this "compassion" which Pellegrino says is "literally" suffering "something of the patient's pain," for there is something at once subtler and more precarious in these words.

Sick people frequently, although not always expressly, want to the doctor to see things from their point of view: 'put yourself in my shoes,' as we say. The patient is surely asking the doctor to do something that the patient regards as vital; what is it? It seems clear that the doctor is *not* being asked to think about the patient's predicament as if it were it fact the doctor's own; the patient's frequent request, 'what would you do if you were me?' is not in the least meant literally, as if the doctor could, 'literally,' become the patient. It is not a matter of some sort of imaginative identification (asking the doctor somehow to become the patient and actually, and *per impossible* to feel the very same pain felt by the patient). Nor is the patient asking the doctor to consider what he would do if he were faced with the same problem, through a sort of imaginative transposal ('suppose you faced this dilemma...'). To 'put yourself in my shoes,' rather, is to ask that the doctor try to appreciate the world in the way the patient experiences things while at the same time remaining himself, this doctor. To be compassionate is not to obliterate the always vital distinction between doctor and patient.

Not to belabor the obvious, the patient is urgently asking the doctor to understand the needful circumstances from his or her own point of view—not literally being the patient but rather playing 'as if' the doctor were the patient. For this, Kleinman insists, the doctor has "to place himself in the lived experience of the patient's illness," to understand the situation "as the patient understands, perceives, and feels it" (Kleinman 1988, p. 232). In different terms used earlier, it is at once both *to understand* and *to be understanding*.

On the face of it, to 'put yourself in my shoes,' may seem extravagant if not absurd. Still, in a sense we do this all the time. Thanks to socially derived and typified everyday knowledge, each of us typically knows something of what it's like, for instance, to be a postman or lawyer, a cook or a sailor, even though we are neither; to drive a truck or a tractor even though we do not drive; to use a wrench or operate a crane even though we have done neither; to suffer acute pain or be faced with an urgent dilemma even though we currently experience neither.

Alfred Schutz pointed out that our everyday knowledge of the life-world is incredibly rich and detailed even while it is also unevenly distributed into different regions (each of us knows some things better than others). Despite the inadequacies, inconsistencies, and inconstancies of commonsense knowledge and understanding, it is for the most part quite sufficient in that context of daily concerns: we 'get along' for all practical purposes. For the most part, when we are asked to 'put yourself in my shoes,' we typically do not go beyond such taken for granted, typified forms of understanding.

At times, though, something more than this typified knowledge and understanding is demanded. The doctor may be urgently asked to appreciate the patient's plight, the dilemma actually faced and experienced: to 'feel-with' the patient from within his or her own perspective and set of moral beliefs, values, and the like. In these terms, putting oneself in the other's shoes involves several critical steps: helping the patient to articulate and understand what that moral framework actually includes, which values the patient has and how these are ordered; identifying which issues seem most pressing, given the patient's concerns, circumstances, and basic ordering of values and commitments; considering the several alternatives with an eye on respective aftermaths and which of them seems most consonant with those beliefs, values, commitments, etc.

To appreciate things from the patient's perspective requires helping the patient to understand and talk about just what this patient believes, desires, aims for, values, and the like—and which most patients have rarely if ever even thought about, making the effort to do so now all the more difficult, in the face of the crisis brought on by illness, decisions, etc. It is in short to invite and encourage a concrete exchange of personal stories, narratives that enable others to enter and become some part of the story.

We are not very often in our daily lives called on to engage seriously in this kind of reflection and self-inspection; when we attempt it, moreover, we quickly realize that it is quite difficult to do and sustain. At the same time, for the professional to provide that kind of help, disciplined self-knowledge by the professional is clearly required; that is, frequently practiced, disciplined reflection intended to delineate the professional's own feelings, moral beliefs, social framework, etc., along with a rigorously discipline suspension of these feelings and beliefs in order to understand what things are like for the other person. This act, a kind of practical distantiation from one's own basic scaffolding of values and commitments, undergirds the act of compassion or affiliation (Zaner 2012).[4]

This act is, it seems to me, vital for establishing an appropriate response to the initiating investment by the patient of trust in the professional. The initiating trust is an appeal to the professional to be responsive to the patient by looking at things from the patients point of view and is elicited in the first place by the professional's own having engaged in that difficult of reflection—more especially the professional's continually engaging in that act together with each patient, to the extent each context demands.

4.12   Care and Dialogue

A significant implication is connected to this set of suggestions. While calling for an "ethics of trust," Pellegrino and Thomasma end up suggesting that "the axiom of care for the vulnerable individual is the ground for an ethics of trust...between doctor and patient" (Pellegrino and Thomasma 1981, p. 185). But they do not apparently appreciate the way in which the patient's act of initiating trust is the ground for such an ethics, not at all the "axiom of care." Indeed, the latter arises as a response to the patient's appeal, itself embedded in that initiating act of trust. In this way, the so-called axiom of care may be called forth, solicited by the pathos at the heart of that initiating act: the vulnerable individual appealing for responsiveness and responsibility within an unavoidably asymmetrical relationship.

Hence, the inner demand of this relation is not only that the professional must never take advantage of the multiply disadvantaged person in need of and appealing for help; it is also that the professional must, by virtue of the commitment to enter a profession (and to remain a professional), specifically 'put myself in the shoes' of each client or patient. This complex reflective act must not only be accomplished by professionals, but among their central tasks is to enable every patient to accomplish that very act for him/herself. It is this complex set of acts that is, I think, at the basis

---

[4] And this is precisely the core sense of what the phenomenological philosophy means by "epoché and reduction".

of professional ethics: in a word, trust and care are mutually, dialectically interrelated in the ethics intrinsic to professional life and are embedded in an ethics that appreciates the issues raised by uncertainty.

Finally, the fiduciary relation is most appropriately understood, then, not as a form of paternalism, but rather on the model of *dialogue*, or perhaps more accurately, on the model of the everyday sharing of *stories*. Precisely because of the complex forms of uncertainty that texture every clinical event, no mater how apparently trivial, communication between physician and patient has vital significance. The key question concerns how that communication should be conceived: whether as a discourse with formalized rules and legal guarantees, or rather as an ongoing experiential discourse of appeal and response, initiating trust and responsive care. Just here, it seems to me, while somewhat misleading on other matters, as I've noted, Pellegrino captures the critical issue: the "moral imperative" that the physician must "be responsive to the way the patient wishes to spend his life." Given that, how should the conversations between doctor and patient be understood? What is the force of this "moral imperative?"

A person needing help asks, appeals to, a doctor for help; the doctor, supposing the appeal is understood, then begins to respond, first by asking questions to which the patient in turn responds. Alternatively, the doctor, so to speak, addresses him/herself to the patient's body and interrogates (palpates, auscultates, etc.), seeking to elicit the body's responses to these 'queries', etc. These questions are followed by further questions and responses, all designed to elicit and progressively delineate what's wrong, then what can be done about it, which decisions seem, together with the patient (where possible; otherwise with those entrusted with decision-making), and eventually what outcomes from which decisions might reasonably be expected, whether one or another of these are consonant with the patient's wishes, values, etc. But the person's appeal and responses to the doctor's 'questions' are not in the least trivial; they arise from and constantly refer to the distress, the dis-ease, experienced by the individual. In one or another way, the patient's appeals and responses are vital for that individual; the more grievous the illness or injury, the more urgent is the appeal.

Thus, for the doctor to be responsive (and assume responsibility) there is nothing for it but to seek in every reasonable and appropriate way for the patient's own ways of experiencing the illness, injury or condition, to probe (on the basis of every clue elicited by and available to the doctor) what the illness, injury or condition means to the patient. In these terms, Cassell is surely correct to emphasize that "all medical care flows through the relationship between physician and patient," and because of that "the spoken language is the most important tool in medicine" (Cassell 1985 I, p. 1). And, the "more" to this, accentuated by Joel Reiser in his Introduction to Cassell's study—that "medical encounters begin with dialogue"—is surely also on target: in the course of the dialogue as here conceived the patient's experiences of illness become transformed into "subjective portraits" or narratives (Cassell 1985 I, p. ix).

To be sure, merely appealing for help does not of itself guarantee any response, much less one that is sensitive to what the patient seeks. In most settings, the doctor to whom the appeal is uttered may well refuse to respond, and for a variety of reasons (themselves subject to assessment, of course). Still, if the doctor does respond to this individual patient, a form of interpersonal relating, a conversation having the form of a certain kind of dialogue (appeal-response, doctor's appeal and patient's response, etc.) then begins and has its own inner demands, course and aims. To respond to the patient means both that the doctor possesses and professes the ability to help, and that whatever the response may be it is inherently open to the patient's further queries—for it is the patient who urgently needs to know and who appeals for help. Not only does the doctor profess the ability to help, but it is essential to the course of the ensuing dialogue that both doctor and patient need to engage each other as truthfully and amply as possible—and for as long as the relationship continues.[5]

To be sure, as with any conversation, the dialogue may break down, for any number of reasons (each of which is, obviously, itself subject to evaluation). In clinical encounters, furthermore, dialogues with patients are intrinsically periodic as well as limited: at various points in the course of diagnosing and treating illness, for instance, the doctor shifts from the person to the embodying organism, in a sense embarking on another kind of dialogue, as noted, to ascertain the nature of the disease process (Pellegrino and Thomasma 1981, p. 112).[6] At some point, too, the doctor's work is over and the patient leaves (permanently or temporarily). Even when such shifts of attention are necessary during the course of the dialogical relationship, however, being responsive to the patient means that the doctor should never lose sight of the specific circumstances of the person being diagnosed and treated, and with whom certain decisions are eventually reached. Nor should the patient's special kind of narrative expressing experiences, feelings, meanings, be overlooked. It is just this characteristic that is central to Rita Charon's emphasis on what she terms narrative medicine (Charon 2006).

The relation between doctor and patient, then, is essentially a special form of what Alfred Schutz termed *Du-Einstellung*: being-oriented-to-another. Schutz points out that this orientation can be either unilateral (as when the other person ignores me) or reciprocal (as when the other is oriented toward me as I orient toward him or her, in this recognizing that I am a person, too) (Schutz and Luckmann 1973 I, pp. 72–88.). In the healing relation, however, the orientation

---

[5] Instances of not telling the truth, factitious illness, Munchhausen's syndrome, and hypochrondriasis, are therefore understandable, and may present the doctor with acute problems, as does the doctor's reluctance or refusal to tell the truth create acute problems for the patient. For the healing relation and its dialogical course begins with the presumption of truthfulness: not to do so—to fake symptoms, or not to inform the patient—is to violate the moral imperatives intrinsic to the relationship. Both can of course be done; both stand condemned, however, in light of those moral imperatives—principally, that which says, respond relevantly and responsibly to the vulnerable patient's appeal.

[6] As noted by these authors, the doctor's part of the dialogue includes a kind of dialogue with the patient's body. In this sense, for instance, palpation is a form of questioning seeking a kind of response from that organism, etc.

to the other is inevitably more intimate, intense, and asymmetrical; but since the very point of the relation is to work toward transcending that inequality, and since the moral imperative at its heart is to be responsive to the person in need of, and who appeals for, help, it is clear to me that this relation is a genuinely mutual one: specifically, a relation embodying and expressing the initiating act of trust (appeal) and care (response). This, it seems to me, is the vital core of dialogue.

References

Cassell, Eric. 1985. *Talking with patients*, vol. 2. Cambridge, MA: MIT Press.

Charon, Rita. 2006. *Narrative medicine: Honoring the stories of illness*. Oxford: Oxford University Press.

Childress, James F. 1970. Who shall live when not all can live? *Soundings* 53: 339–355.

Childress, James F. 1982. *Who should decide? Paternalism in health care*. New York: Oxford University Press.

Engelhardt Jr., H.T. 1986. *The foundations of bioethics*. New York: Oxford University Press.

Faden, Ruth R., and Tom L. Beauchamp. 1986. *A history and theory of informed consent*. New York: Oxford University Press.

Gadow, Sally. 1981. Body and self: A dialectic. In *The humanity of the ill: Phenomenological perspectives*, ed. V. Kestenbaum. Knoxville: University of Tennessee Press.

Hardy, Robert C. 1978. *Sick: How people feel about being sick and what they think of those who care for them*. Chicago: Teach'em, Inc.

Hippocrates. 1923. *Epidemics I*. Trans. W.H.S. Jones. Cambridge, MA: Harvard University Press, Loeb Classical Library.

Kleinman, Arthur. 1988. *The illness narratives: Suffering, healing and the human condition*. New York: Basic Books.

Lenrow, Peter B. 1982. The work of helping strangers. In *Things that matter: Influences on helping relationships*, ed. H. Rubenstein and M.H. Bloch, 42–57. New York: Macmillan Publishing Co.

Liddle, Grant. 1967. The mores of clinical investigation. *Journal of Clinical Investigation* 46: 1028–1030.

Marcel, Gabriel. 1940. *Du refus à l'invocation*. Paris: Gallimard.

Pellegrino, Edmund D. 1979. The anatomy of clinical judgments: Some notes on right reason and right action. In *Clinical judgment: A critical appraisal*, ed. H.T. Engelhardt Jr., S.F. Spicker, and B. Towers, 169–194. Boston/Dordrecht: D. Reidel Publishing Co.

Pellegrino, Edmund D. 1981. Being ill and being healed. In *The humanity of the ill: Phenomenological perspectives*, ed. V. Kestenbaum. Knoxville: University of Tennessee Press.

Pellegrino, Edmund D. 1983. The healing relationship: The architectonics of clinical medicine. In *The clinical encounter: The moral fabric of the patient-physician relationship*, ed. E.E. Shelp. Dordrecht/Boston/Lancaster: D. Reidel Publishing Co.

Pellegrino, Edmund D., and David Thomasma. 1981. *A philosophical basis of medical practice: Toward a philosophy and ethic of the healing professions*. New York/Oxford: Oxford University Press.

Pellegrino, Edmund D., et al. (eds.). 1991. *Ethics, trust, and the professions: Philosophical and cultural aspects*. Washington, DC: Georgetown University Press.

Schutz, Alfred. 1964. In *Collected papers*, vol. II, ed. Arvid Brodersen, 91–105. The Hague: Martinus Nijhoff.

Schutz, Alfred, and Thomas Luckmann. I: 1973. *Structures of the life-world*, 2 vols. Evanston: Northwestern University Press.

Sokolowski, Robert. 1991. The concept of professions. In *Ethics, trust, and the professions: Philosophical and cultural aspects*, ed. E.D. Pellegrino, et al. Washington, DC: Georgetown University Press.

Spiegelberg, Herbert. 1986. *Steppingstones toward an ethics for fellow existers: Essays 1944–1983*. The Hague: Martinus Nijhoff.

Zaner, R.M. 1994 OP. *Troubled voices*. Cleveland: Pilgrim Press.

Zaner, R.M. 2012. *At play in the field of possibles: An essay on free-phantasy method and the foundation of self*. Bucharest: Zeta Books.

Chapter 5
Openings into Clinical Ethics

5.1 What Are We to Make of the Backlash Against Ethics?

I want to pick up where I left off in Chap. 3. In some respects, it is peculiar that, in the face of the resounding backlash against 'Big Ethics,' as it was often called in the 1970s, some physicians continued to entertain the notion that philosophers should, and some of them argued must, become "involved" in clinical medicine. Around the same time (early 1980s) as Alan Fleischman was putting his program for residents in place, for instance, the pediatrician Tomas Silber stated his belief that without such actual, regular involvement in clinical affairs, what he termed the "data base" for understanding, much less contending productively with, the moral issues he regarded as inherent to the daily practice of at least pediatric medicine, would be plainly missing. Precisely that "base" is necessary, he argued, for the medical tasks at hand in any clinical situation. Thus, quite understandably, Silber lamented the "absence of these professionals"—that is, philosophers—"from our daily lives," although, with Siegler, he endorsed the idea that physicians must for their part immerse themselves in philosophy and theology (Silber 1981).

In the meantime, the person who has been, by any estimate, the dean of this entire enterprise, Edmund Pellegrino, was already eyeing much larger horizons. He had many times stated his belief that the times and the issues are right for a "new Paideia" matching that of classical Greek culture, and that medicine and philosophy occupy the pivotal places in that endeavor now as they did then (Pellegrino 1979). He argued for many years that medicine, as he said, is the most scientific of the humanities and the most humanistic of the sciences, and presumably the interstice thereby generated is precisely where both must take up residence from which to cultivate, along with philosophy and others of the humanities, that new Paideia. Even more, he argued that a proper understanding of some of philosophy's own perennial issues positively requires a sound grasp of what medicine has learned— for instance, about neuronal activity or the human body more broadly.

© Springer International Publishing Switzerland 2015
R.M. Zaner, *A Critical Examination of Ethics in Health Care and Biomedical Research*, International Library of Ethics, Law, and the New Medicine 60,
DOI 10.1007/978-3-319-18332-9_5

For that new Paideia to emerge, however, philosophers and physicians must probe with far more sensitivity and depth than hitherto their common as well as distinctive methods and problems. What divides them has too long been too divisive and antagonistic—without, moreover, the least justification, as would be clear from even a cursory glance at their long histories. While it is true that some philosophers early on took exception to the notion that ethics in medicine is merely an 'application' of ethical theories to practical medical problems, (Caplan 1982; Toulmin 1982; MacIntyre 1981; Gorovitz 1982) the idea of that "new Paideia" seems as far from realization today as it did in the 1960s and 1970s when Pellegrino first proposed it. That notion, in any event, has long seemed to me very much worth pursuing; in fact, much of my writing over the past several decades, moreover, has been devoted precisely to it.

Here I propose to pick up that theme once again and make it the predominant one. This will inevitably bring me into some rather novel thematic directions—specifically, probing the quite new prospect of philosophy within the context of clinical work. I have long been convinced that there is something very important in the idea of that clinical involvement. The discipline it imposes, furthermore, will just as inevitably raise our sights onto very different vistas, to themes that may seem quite strange to our accustomed ways. In what follows, that is what I propose to pursue; and even when there are other themes which will preoccupy me from time to time, I want to make it clear that my over-riding concern will be to test those clinical waters for the ability of philosophers not to sink or become distorted beyond recognition. And, to repeat, it is the idea of that new Paideia that will serve as the guide and goal of this inquiry.

For that undertaking, ethics has to be understood as a preeminently practical discipline and at the same time one that is ingredient in all its components in philosophy; hence, philosophy itself has a serious commitment to the issues and the life of praxis. Aristotle, I believe, was correct (Aristotle 1962). Pointing out in his Nicomachean Ethics that "precision cannot be expected in the treatment of all subjects alike" (I, 3), he understood that "when the subject and basis of a discussion consist of matters that hold good only as a general rule, but not always," as is the case with politics and ethics, "the conclusions reached must be of the same order" (I, 3). So far as moral actions are concerned, "although general statements have a wider application, statements on particular points have more truth in them: actions are concerned with particulars and our statements must harmonize with them" (II, 7). Concerned with emotions and actions in the practical realms of life, neither the study nor even the discussion of ethics permits "the kind of clarity and precision attainable in theoretical knowledge" (II, 3).

In ethics "there are no fixed data in matters concerning and questions of what is beneficial, any more than there are in matters of health" (II, 2). The treatment of particular problems will be even more characteristically imprecise; therefore, here, "the agent must consider on each different occasion what the situation demands, just as in medicine and in navigation" (II, 2). It is my conviction that, at least as far as these citations are concerned, Aristotle's view is correct.

5.2 Several Views of Clinical Ethics

Following on the brief overview of medicine in Chap. 2, it is necessary to give the same sort of overview of ethics as a clinical discipline. Specifically, we need to look into the several proposals that have been offered about that. This will lead to the expression of certain clinical ethics theses as a summary way of expressing the sense of such a discipline, and will be developed in much greater detail in subsequent Chapters (Zaner 1994).

5.2.1 The Clinician

In several splendid essays, John La Puma and others (La Puma 1987; La Puma et al. 1988; La Puma and Toulmin 1989; Schiedermayer et al. 1989) have defended the view that there is a legitimate role for the ethicist working within the clinical context as a consultant. Mentioning several models for this role, they argue that only a clinical model is appropriate. Possessing special knowledge and skills, as these authors see it, the ethicist should be responsible for helping physicians become more sensitive to ethical issues. In this, the clinical ethicist is precisely like any other clinical consultant, bringing special knowledge and skills to bear on special problems flagged by an attending physician as needing that expertise—most often, as these occur as regards some specific patient. I should note here, in passing, that there is hardly a mention of other health professionals, patients with their significant others, or the social nexus of the practice of medicine.

Although they pose the idea that the ethicist could be either physician or non-physician, they in fact argue that the ethicist must be able to interview and do physical examinations of patients—as well as speak with families, discuss cases with members of the health care team, review charts, and document recommendations in patient's medical records. In short, he/she must be an experienced clinician able to help the attending manage patients—which suggests that the physician is the one who is best able to serve this role, as Siegler argued a long time ago (Siegler 1979). For this view of ethics, the ethicist's credibility and effectiveness depends not only on knowing ethics but, of equal if not more importance, on possessing a fund of relevant medical knowledge, hands-on clinical patient care skills, and the ability to discern medical distinctions that are technically or morally relevant in caring for patients (i.e. to make clinical judgments, and to be as accountable for them as any other clinician). In fact, therefore, this view of ethics proposes that (a) only a clinically experienced physician can serve as a clinical ethics consultant, and (b) such a consultant serves the attending physician first and foremost, if not exclusively (La Puma and Priest 1992; Edwards and Tolle 1992).

5.2.2   The "Expert"

Except for the clear implication that physicians can best serve in this role, George Agich (1990) had already stated and expanded on that proposal some years ago. He delineated four roles inherent to the idea of clinical ethics: teaching, watching, witnessing, and consulting. Moreover, from his comments on my work, it is clear that Agich still endorses his basic view set out in the earlier article (Agich 2005). Clinical ethics is different from more traditional forms of teaching, as it is conducted in the practical settings of clinical medicine. It is, moreover, a form of practice similar to other forms of clinical practice, although it is more involved in role modeling, character building, and skill acquisition than with theoretical, cognitive understanding.

Watching, in Agich's usage, involves the methodical, disinterested, and objective study of medical practice. *Witnessing*, by contrast, draws one into the scene of social action as an agent who is accepted by others also involved; the witness is thus a resident expert who is able to provide useful, practical advice to those others. Appealing to an anthropological model derived from the work of Charles Bosk (1985) Agich sees witnessing as most important. As a witness, the ethicist is "more directed to establishing or ratifying a moral community than mere watching..." (Bosk 1985). Somewhat like a priest invited into the private meditation and ceremonies of the group—where inmost secrets, uncertainties, and anxieties are revealed and shared—the witness comes "to symbolize for the group the moral community outside the hospital..." (Bosk 1985). He also insists that this role includes helping to give social definition to clinical realities, and may even provide subjects with a sense of legal protection, although it is unclear in just which sense this "protection" is understood—whether as a form of 'cover your ass,' or, more unlikely, I suppose, as giving actual legal information and even advice (in which case, however, the ethics consultant must also then be a licensed attorney).

One natural outgrowth of this unique form of participation is *consulting*, the fourth activity of clinical ethics. As for any clinical consultation in clinical medicine, this special ethics activity naturally carries expectations of practical help in decision-making. It thus functions strictly under the aegis of the attending physician and within established institutional rules and procedures. The consultant may be an independent professional, but the work of anybody with this role is established and remains under the aegis of the attending physician, whose work is thus directed to problems perceived by the physician and associated with the primary physician-patient relationship. Such a consultant is assumed to have relevant expertise, skill, and training to identify and evaluate the issues involved in caring for an individual patient (Agich 1990, p. 392).

The expectation of practical help is central. The ethicist is expected to possess and utilize the relevant expertise, skills and methods in order accomplish the consultative goal. While not necessarily a physician, as Agich sees it, the ethics consultant is nevertheless a specialized clinical practitioner "who brings an independent expertise to bear on problematic cases or issues in medical practice," and who functions in clinical situations "as an identified expert...to offer advice or recommendations in specific cases presenting ethical problems" (Agich 1990, p. 393).

5.2.3 The Casuist

But with respect to what is the ethicist a so-called 'expert'? In an early article focused on just those situations that might reasonably prompt an appeal for a specialist in ethics, Albert Jonsen raised several critical questions that are pertinent to this question: "What sort of knowing...is ethics? Above all, what is its use?" (Jonsen 1980, p. 158). He went on to propose a way of approaching and thinking about these issues within clinical settings. Although it took some time for him to identify just this "way," a few years later he then more confidently declared, "hospital philosophers or ethicists are, in fact, casuists" (Jonsen 1986).

For Jonsen, the usual approaches to ethics are not helpful, for they miss the main point and focus on decision-making: not conceptual analysis of issues, but direct involvement in the decision making process. Far from a novel proposal, he points out, this is an idea with a long history in western ethical traditions, specifically in casuistry. This approach in general involves considering morally perplexing cases "in the light of certain general ethical norms or rules," that is, where "a definite view of the nature of the moral life is confronted with a well-described real or fictional situation." As such discussions call for a specific and practical response, not to abstract concerns but to the particular set of circumstances under consideration, they invoke a clear focus on the question, "What should be done in this situation?" (Jonsen 1980, p. 159)

For Jonsen, then, casuistry is not merely one more way of applying principles already at hand to particular clinical cases, since it is only when a number of such cases are arrayed together that the notion of "principle" itself begins to have significance. In the practice of casuistry, properly understood, a comprehensive ethical theory does not precede but follows the study of particular cases—though it is surely true that moral problems are constructed around "an already perceived but, as yet, inarticulate moral notion" exemplified by particular cases (Jonsen 1980, pp. 169–171).

This focus on the decision process suggests that casuistry and clinical ethics consultation are very much the same activity: both are forms of reasoning directed toward practical resolutions that lead to decisions, and from there to practical actions. In these situations, uncertainty and probability, not "the truth," are the centering themes. Both have the aim of assurance or, as the casuists said, a "'certain conscience';" that is, Jonsen explains, "resolving practical moral doubt" (Jonsen 1980, p. 163). In fact, he insists, casuistry did not presuppose a shared worldview but came about mainly during times of social fragmentation; it "thrived on doubts, uncertainties, and dilemmas, and moved toward the creation of an ethic rather than from one already formed," and precisely here was the place of "assurance."

Working in situations where an explicit moral consensus is missing and only probable opinions can be offered, the casuist-consultant methodically follows three steps: typification,[1] use of maxims, and assurance. A 'case' of interest to a casuist will present a typical moral dilemma that can be understood only within specific circumstances, which include some moral virtue (justice, charity) or rule ('Thou shalt not kill,' for example), and actors with certain social roles (parent, priest, proprietor) who encounter each other within specific social relationships (contracts, promises, requests). The case is a typification or, in Jonsen's preferred term, paradigm[2]—a situation that is neither wholly unique nor fully universal, but rather is textured with "roles and relationships that are morally relevant to its interpretation and resolution" (Jonsen 1980, p. 165). As I prefer to say, it is a context.

Built around some virtue that is usually not well-articulated, the case also typically includes a reference to moral maxims called "reflex principles"—statements, the truth of which could not be completely demonstrated but were commonly accepted as having to be weighed in moral deliberations. Not taken for granted in the casuist's discussions, they had instead to be tested or interpreted for their pertinence to the problem at hand—like "shuttles that move back and forth within the texture of the roles and relationships in order that a pattern can appear" (Jonsen 1980, p. 166).

The search for assurance follows quite naturally. Directly confronting the inherent uncertainty of every case, the casuist mainly sought to provide assurances to the moral agent that an action which provoked some apparently un-resolvable doubts, could nevertheless be performed with practical moral certainty. As Jonsen insists, however, "the 'new' casuistry must be more than talking about cases. It must be an articulated art, that is, it must be able to discuss the singular and unique in terms that can be generally understood and appreciated. It must have the quality of moral discourse" (Jonsen 1986, p. 71).

5.2.4   The Facilitator

The first two views contend that the ethics consultant should be conceived on the medical model of independent expert brought in by the physician in charge, while Jonsen's view is that 'problems' faced by any agent are the central occasion for the casuist's thinking, and that the isomorphism between consultation and casuistry apparently holds only for ethics, not medicine; it remains unclear, however, how such an approach is either 'clinical' or how it could become involved in clinical

---

[1] Jonsen refers to the work of Alfred Schutz as highly significant for this crucial concept. It is, of course, Schutz's native air (Schutz and Luckmann 1973). In their book on casuistry, Jonsen and Stephen Toulmin present a more detailed methodology (Jonsen and Toulmin 1988, pp. 307–14).

[2] This switch from "typification" to "paradigm" obscures precisely what Jonsen otherwise wants to emphasize: the common uncertainties ingredient to such situations as attract his attention—which 'paradigm' hardly makes patent.

encounters. It would seem that such involvement could occur solely at the request of, and mainly for the benefit of, the physician; access, if you will, to the patient could occur only with the physician's specific permission. And, still, it is worth noting here, as earlier, that there is not a word about the array of other health professionals whose work is clearly highly significant for patient care: nurses, specialist technicians, and the like.

The idea that ethics consultation is different from what the medical model requires is also proposed by others, among them Glover, Ozar, and Thomasma, who specifically wish to differentiate the "individual expert" model that has been an important part of medicine from another which they believe is more appropriate for clinical ethics: the "decision facilitator" (Glover et al. 1986).

In the first model, when a physician needs advice about a particular aspect of patient care, he or she calls on a specialist in that area to provide an expert judgment. The consultant is not expected to take actions to insure that decisions are made, nor even to participate in that process. Both consultant and attending "function as relatively isolated individuals" (Glover et al. 1986, pp. 22–23). The ethicist is interpreted as an "expert" who, like the medical consultant, has specialized knowledge necessary to resolve the special ethics issues in patient care.

The difficulty with this view, Glover et al contend, is that the assumption of shared knowledge, training, methods, and values doesn't hold up. There simply is no consensus of the sort needed to make the idea of "expert" coherent; there is no way for the physician asking for the consult to know how to assess the ethics "expert's" views. The problem concerns not so much the kind or adequacy of training, but the nature of ethical decision-making itself—which, in the end, is much less an individual and much more a community undertaking, and is thus different from the medical consultant.

They propose instead that the ethicist serves as a "facilitator" of decisions, and in this sense acts to convene a group of knowledgeable individuals and then promotes a discussion designed to sort out the essential information needed for understanding issues and recommending decisions. What makes these clinical discussions specifically ethical, then and quite unlike Jonsen's "casuistry" model, stems from doing precisely what Jonsen denies: "applying" so-called ethical principles to specific cases—although, again, the very idea of "application" is left completely unexamined. The model thus endorses the commonly received "principles" view of ethics, which is preferred "because ethical wisdom and sound decision-making are at their best when they can draw on the perspectives and insights of a community of persons rather than a single individual," (Glover et al. 1986, p. 24) and thus permit consideration of a greater range of values and interests, and promoting more effective communication and action.

Although the medical model may make some sense for physicians, for this proposal about clinical ethics, it is not appropriate for ethics or ethical decision-making—always a communal endeavor requiring consideration of a far wider range of issues than is possible in the "expert" model.

It is important to note that subsequent writings on ethics consultation have continued to follow one or another of these models—most often emphasizing the centrality of the attending physician for any discussion of ethical issues, from whatever perspective. For instance, the American Medical Association's Code of Medical Ethics (especially "Opinion 9.115—Ethics Consultation") was published in 1998 following the adoption in 1997 of that Code. This opinion follows what others have also come to accept: namely, that ethics consultation should be conducted as part of the work of an Ethics Committee. And, notably, a lengthy, multi-authored survey article on ethics consultation in 2006 (Orlowski et al. 2006) did not go beyond any of the models cited above, and is based not on actual efforts at such consulting but on a survey among ethics committees.

5.3 Ethics in the Clinical Encounter

Clearly, there is much that is problematic about clinical ethics: what is or should be done under this aegis? What are reasonable expectations (by patients/families, physicians, and others)? Who should consult? What methods are to be utilized?—indeed, whether the very idea is at all legitimate and thus worth pursuing remains in question. At this point, it is well to assess each of the proposed views.

5.3.1 The Clinician

The proposal that the ethicist should function precisely like any medical consultant risks confusing both ethics and medicine. While it may not be wrong to regard an ethicist as having a certain body of special knowledge and experience, the main issue concerns the details of that role. To be sure, it may be readily conceded that the ethics consultant should be capable of functioning in clinical situations. However that may be, it seems to me that La Puma confuses matters, specifically in the claim that the ethicist should not only talk with the physician who requests the consultation, but in addition and *as ethicist* must be able "to examine" patients and help in their medical management. To ask for this kind of involvement is to say that the ethicist must be a physician. It is surely unreasonable, and probably illegal, to expect a non-physician to have those skills and perform those actions—which is obviously not to suggest that the ethicist can not be an experienced clinician and held accountable for whatever is done under that aegis.

There is another problem with this view, however. Not only is it left unexplained just why an ethicist must also be able to conduct physical examinations, so it is left unexamined why a physician must also be able to conduct ethics examinations—as if a clinical-ethics examination were precisely like a clinical-medical examination. While there may very well be some similarities between clinical methods followed by the examining physician, and clinical methods followed by an ethicist, they are surely substantially different: to know how to detect and assess an irregular heartbeat, for instance, does not in the least provide insight into why the same patient is morally troubled nor in what those 'troubles' consist. The heart of the difficulty here, as will be explored later on, may well lie in a too narrow understanding of what 'clinical' signifies. After all, the clinical activities of, say, a nurse, a nurse practitioner, an EEG technician, or a social worker, do in fact differ significantly from those of a physician—but are widely accepted as 'clinical' quite as much as the doctor's work. That there are similarities is clear, too, although it will take some analysis to clarify both the differences and the similarities.

It should be noted, furthermore, that neither Agich nor La Puma says anything about the need for ethicists to talk with nurses and other health professional staffs. Even more striking, neither has much to say about why and how ethicists should be able to work with patients and their families and/or significant others, much less in what such 'work' should consist—a shortcoming of most of the writings about clinical ethics as well as ethics committees.

In some ways, nonetheless, Agich's position does not present much with which to quarrel regarding teaching or watching. Not even witnessing is all that troubling—though the idea that the ethicist evokes the image of 'priest' is not only unsettling but seriously erroneous—no ethics consultant has to work from within a taken for granted religious tradition, for instance. It may also be true that when invited into the inner sanctum of clinical practice, the ethicist's function may be regarded as a sort of 'ratifying' or even 'establishing' of a moral community— although that, too, it seems to me, is at the very least elliptical. It is not that there always is some sort of taken for granted 'moral community' shared by everyone involved in some clinical encounter—indeed, it may well be that the absence of such a commonality constitutes some of the very issues that need attention. Still, it may also well be correct to point out that the very presence of an ethicist in a particular encounter does indeed signify that attention to moral issues is not only possible and important, but is also imperative.

It is nevertheless peculiar at best to propose that the ethicist "symbolizes" the broader "moral community outside the hospital." Indeed, should that occur (as may perhaps at times happen), it can be very troubling precisely because it can compromise the point of an ethics consultation—to focus on the specific circumstances and identifiable issues inherent to the individual case in question. One need only consider one expression of that external moral community, for instance the deep disputes over stem-cell research and the variety of proposed regulations, to make it vividly clear that many ethicists feel as much tension and conflict with the outside community as may be found within the health care community.

All things considered, it seems to me, the philosopher who gets involved in clinical settings more often serves as a reminder to health professionals that their interventions must be placed within a far richer context of interrelationships than has been usual, especially as those interventions may well have been the actual source of the issues that provoked the consult in the first place. Insisting on the need to address ethics issues serves as well to remind health professionals of the moral complexities and uncertainties inherent even to routine clinical situations—that statistical information, for instance, can be more confusing than helpful for many patients and their families. But this is not at all anything like a priestly role. The ethicist is not a representative (much less conscience) of the outside community but rather a reminder of the common humanity of all those involved in the case, and embodies the insistence that deliberation about the moral dimensions of any clinical event is quite as significant as attending to bodily problems—and, I must add, in no way confused with the need of some (but only some) patients to have religious needs and feelings appreciated.

However that may be, it is Agich's apparent endorsement of the "individual expert" model that is the most problematic. First, the claim is troublesome even if the ethicist is directed to the primary physician-patient relationship; possesses knowledge, skills, and training; and works within established institutional rules and procedures. Glover et al are quite correct to emphasize the inherent problems about such "expertise." Still, there is little that is commonly accepted about what 'knowledge,' which 'skills,' or what 'training' the ethicist needs—although, as Jonsen says, there is some consensus about the issues presented by some of the well-known so-called paradigms (Jonsen 1986, p. 72).

Beyond the many problems hidden in the idea of 'the expert,' as I'll point out in a later Chapter, the ethicist does not and cannot function in isolation—an issue that will be taken up shortly. It is enough to note here, as I have emphasized earlier, that it is precisely because of the ethicist's focus on the clinical encounter that the work of ethics consultation is necessarily complex and cannot be done alone. Not only is that relationship set within an immediate complex of people (patients, physicians, nurses, social workers, chaplains, family, friends, and still others), each of whom has, or claims to have, something at stake that cannot be ignored in clinical decisions. There are also numerous types of written and unwritten norms, rules, procedures, regulations, standards, etc. that help to shape and govern what goes on—and cannot, therefore, be ignored.

To understand what problems are presented in that encounter, therefore, necessarily brings the ethicist into relationships with a network of other people, both directly and indirectly present in and part of every encounter. Merely to illustrate: there are not only relations between physician and patient, but among physician and family members, family members and the patient, nurse and physician, nurse and patient, nurse and family, to mention only a few. To ignore any of these may well result in ignoring or obscuring precisely what is ethically at stake in a given situation.

All of these must be recognized, appropriately assessed, and dealt with in oftentimes quite different ways. To do so, all these stakeholders—or least their legitimate concerns—must be as much a part of 'ethics discussions' as the patient or physician. The ethicist's role, I think, is better conceived as convener and facilitator than as individual expert working independently. This complex context of people, standards, expectations, etc.—i.e. relationships constitutive of the clinical encounter—clearly also plays its part in the physician's work; hence, not even the physician, much less medical consultants, can proceed in ignorance of that context of multiple relationships. These considerations raise a good many issues that will be best addressed on their own.

5.4 The Casuist

The most intriguing discussions of clinical ethics are provided by Jonsen and Toulmin, and by Glover et al. As will be clear shortly, my understanding of this discipline converges with theirs in several ways while departing in others. For now, however, let me only note several problems.

Jonsen and Toulmin's reminder in their book (Jonsen and Toulmin 1988) of what lies in our own moral traditions is immensely important. Still, several matters give pause. On the one hand, while I agree that the focus of clinical ethics must surely include decisions that are made or need to be made, what lies in the background of this focus for these authors is troubling: that discussions of cases go on "in the light of certain general ethical norms or rules," in Jonsen's words. To be sure, this does not require anything like a full set of ethical principles; casuists "did not apply principles to cases in any carefully deductive or inferential fashion" (Jonsen 1986, p. 68).

It is also true that the casuist's concrete method is far richer than is portrayed in either the 'applied' or 'principles' model of ethics. Still, the distinction between the casuist and the 'application' model at times seems only a matter of emphasis. Whether a case discussion is constructed around an ethical principle or only "an already perceived but, as yet, inarticulate moral notion," the point remains the same: namely, in some sense (most often unspecified) 'to apply' the principle or the moral notion to the particular circumstances, in which case the alleged difference seems all but to disappear.[3]

However that may be, the casuist position remains unclear at several critical points. First, *whose* "doubts," "uncertainty," "typifications," "maxims," etc. set the agenda of the moral discourse in clinical situations? *Whose* concerns are at issue and why? *Whose* decisions are to be noted and analyzed? Among clearly available and at times quite different "paradigms," what justifies the selection of one over another? The physician's? The nurse's? The patient's? One or another member of the patient's family? Which and why? The medical consultant? The casuist-consultant's? In the end, we never know just whose "perceived but, as yet, inarticulate moral notion" sets the agenda for the discussion in the first place, especially in clinical encounters. Rather than facing one or even two such perceptions and notions, I suggest, the consultant in ethics more commonly confronts a veritable chorus of them—not always (indeed, rarely) in harmony (Zaner 1993a, b).

It therefore remains seriously unclear with whom the casuist-consultant is supposed to consult, much less about which ethical issues one is supposed to

---

[3] In Jonsen and Toulmin's study, this point is even clearer. For instance, they state that the "first substantive task [of casuistry] is to agree just which 'paradigm' best fits the circumstances in question" (Jonsen and Toulmin 1988, p. 308). By "fit" it is clear they mean "apply," as is stated explicitly a bit later: "Deciding what type case, or paradigm, best applies in any given circumstance," even though this is said to be "only the first step..." (pp. 311, 312). My point is merely that it is quite difficult to see much difference between the casuist and the applied ethics models their argument for a difference seems a matter of smoke and mirrors.

address, not to mention how one settles a dispute over just what is at issue. Jonsen's view seems to be that when meeting with the physician (he does not mention meeting with a patient/family, nurse, etc.), the consultant may suspect and then perhaps detect some "as yet inarticulate moral notion(s)" and "perceptions" at work. Then what? I suppose that the ethics consultant is then supposed make these clear in the light of and as 'applied' to the actual clinical circumstances (which have, one must hope and assume, been appropriately probed and understood), and then to assist the physician to reach an understanding and critical assessment of options and outcomes in order to allow needed decisions to be made in the best possible manner.

Apparently, however, none of this work is done for the sake of anyone besides the physician—not the patient, not any family member or other loved ones, nor the nurse, social worker, etc. At the very least, moreover, this process itself—its method, steps, and whatever is required by that 'detective' work—remains significantly obscure: just what, precisely, does such a casuist consultant actually *do* in order to detect some "as yet inarticulate moral notion"? What is actually done, what are the details of the 'method' actually are, remain profoundly unclear.

Second, there are several quite significant issues that are suggested but not well addressed—and these may well be decisive. To be clear about the notion of casuistry, it is crucial to understand just whether and how the casuist knows *which* "as yet inarticulate notion" is at issue in any specific case, and knows moreover that the perceptions and notions are not brought in (silently or unwittingly) by the consultant or someone else. How, after all, does one go about bringing into the open forum of discussion notions that are inarticulate and obscure? Not that such knowledge is impossible; indeed, it may well be precisely part of what needs to occur in an ethics consultation. My point is rather that the consultant's claims to knowledge require *evidence* as well as serious clarity about the details of the so-called 'method.' As far I can tell, however, the casuist's proposal presupposes an entire theory of interpersonal relations (Zaner 1981) and communication, not to mention a more ample methodology, before it can even be adequately assessed.

Everything said about casuistry could well be true, that is, and we would still not know much about the consultant's actual role, methods, or goals in any specific situation. This is crucial: after all, the "most important" thing about casuistry, these authors allege, is the recognition that uncertainty affects every particular case. Since these situations often involve having to weigh and then just as often to reach irreversible decisions based on that uncertainty, "assurance" or "conscience" has an equally central place. But what is not said is just how that weighing is to be done, *specifically and concretely*, especially regarding the specific forms of evidence at hand on which to base this critical process.

The more perceptions and 'moral notions' each case presents, the more complicated are the issues facing the consultant. For instance, while the physician may well give weight to a certain maxim and course of action, the patient may find that action unacceptable even while accepting the maxim—or vice versa (Jonsen and Toulmin 1988, pp. 16–19). Similarly, while a patient's family may 'perceive' a situation one way, other decisional stake-holders may perceive and understand it quite differently. In one clinical event, for instance, when a patient refused dialysis and the family wanted it instituted, the renal consultant was uncertain whether it would help and was thus uncertain about what to do, the attending wanted merely to avoid conflict, the nurse thought the patient was being 'flogged' unnecessarily, and the hospital attorney insisted on the need to avoid possible legal suit (Zaner 1993c, pp. 47–56). In another encounter, while the attending was convinced that further treatment was futile (and several consultants agreed), the woman's daughter and son cancelled the DNR order after they had at first agreed. Discussion revealed that while her middle-aged children understood very well that CPR would be inappropriate, they were concerned over the range of the DNR order: would it mean that no anti-biotics would be used if their mother became septic again? In such situations—certainly very common in hospitals today—it is difficult to figure out just how the casuist model works. In fact, it may work best only in relatively uncomplicated clinical cases; or, indeed, only where there is no disagreement among various parties. It grows increasingly unclear when the case is fleshed out with the multiple voices that are invariably presented.

In their more detailed study, Jonsen and Toulmin do recognize one facet of the difficulty. Similar cases or paradigms, they contend, are the final objects of reference in moral argument, and carry certain presumptions and weight unless exceptional circumstances are present. The first task with any particular case is "to decide which paradigms are directly relevant to the issues each raises" (Jonsen and Toulmin 1988, p. 307). Difficulties arise when either the paradigms fit only ambiguously and the presumptions are open to challenge, or if two or more paradigms apply in conflicting ways (and thus must be mediated)—and this suggests the problem.

They consider an obstetrician's dilemma when faced with an ectopic pregnancy (Jonsen and Toulmin 1988, pp. 312–13). Here, the physician faces two professional duties: to preserve the mother's life, and to deliver the infant unharmed. As the pregnancy could be fatal to the mother and cannot yield a viable infant—the ectopic implantation is a fatal disorder for the fetus—the physician operates to remove that part of the fallopian tube where the fetus has implanted, thereby ensuring that the fetus will die sooner rather than a short time later. Although needed details are lacking in this case, there is little reason to quarrel with the way they construe the issue thus far, nor with other important matters they note (such as that this is merely one type of moral conflict). What is troubling, rather, is that such matters are invariably seen from the perspective of the physician—surely not the sole, at times not even the weightiest of perspectives—as if, for unexplained reasons, the physician enjoys a morally privileged status, which is obviously dubious in this case.

Particular cases present a multiplicity of voices not always in harmony—each of which, although rarely mentioned, could doubtless lay claim to having been poorly considered, or not at all, in the casuist's terms. In these cases, however, the first and continuing issue is none of those Jonsen and Toulmin enumerate, at least not simply those. The concern is or ought instead to be, not only to enable known decisional

stake-holders to have their fair say in these matters, but at times to discover just who
are the appropriate stake-holders and their relative weights—which requires determining, within the practical constraints of circumstance, how best to arrange time
and place so as to allow them to be heard. Jonsen and Toulmin's move to the
"paradigmatic" or the "analogical," in a word, strikes me as quite hasty; they seem
overly anxious to reach knowledge (*episteme*) and as quickly as possible to leave
the domain of action (*phronesis*)—a move that is at best risky, as it ignores the very
thing that makes the clinical situation what it is all about in moral terms.

At one point, Jonsen and Toulmin, however, are quite right:

> The heart of moral experience does not lie in a mastery of general rules and theoretical
> principles…It is located, rather, in the wisdom that comes from seeing how the ideas
> behind those rules work out in the course of people's lives…. In ethics as in medicine, this
> "practical experience" is as much collective as personal. (Jonsen and Toulmin 1988, p. 314)

Having said this, however, in explaining what they mean by "collective," they
emphasize that the ways of resolving conflicts, working out priorities, settling
ambiguities, etc., are fundamentally products of "the long and rich histories" that
have been woefully neglected. To note such neglect is not irrelevant, and perhaps is
even correct in its way, but it is plainly unhelpful regarding the feature of clinical
cases at issue here—not history so much (which fascinates us philosophers, but
rarely patients or physicians) but *sociality* must be recognized; not 'others' in our
historical past so much as those actually involved in each individual case—where
those multiple voices, each compelling respect, at times quite loudly but at others
muted, constitute what in large part the case is all about.

In short, whatever else may be said, Jonsen and Toulmin simply ignore a number
of essential issues: those regarding the pregnant woman and her husband (about
whom we are told nothing), the developing fetus and its moral status, other siblings
(if any), the possible problems one or another nurse might have, and still others. Nor
is mentioned made of any of the complex of institutional, governmental, or social
contexts of rules and regulations that clearly variably frame and impact such
situations.

5.5   The Facilitator

In a way, Glover et al recognize the point I am anxious to make. The main thrust of
their criticism of the "individual expert" model is its insistence that ethics consultants, like medical consultants, are taken as solitary individuals who possess some
sort of special knowledge. As opposed to this, they insist that the ethicist be
construed as a "decision facilitator"—explicitly recognizing, as opposed to the
casuists, that others are necessarily involved and have their respective voices,
stakes, and knowledge pertinent to decision making. The role of facilitator is to
bring together appropriate individuals to sort out the essential information needed
to provide understanding and make sound decisions. The reason for this is that

"ethical wisdom and sound decision-making are at their best when they can draw on the perspectives and insights of a community of persons rather than a single individual" (Glover et al. 1986, p. 24). Conceiving decision-making in this communal way is said to permit consideration of a greater range of values and interests, insure better and more effective communication, and be more likely to result in justifiable action.

Although I agree in a way with this emphasis on "a community of persons," several difficulties are worth noting. First, it should at least be mentioned that it is quite unclear just what is to be understood by "appropriate community"—what, after all, determines what 'appropriate' means, and who should be included and why, or whether just any 'community' will do as well as any other—much less precisely why "ethical wisdom and sound decision-making" are allegedly "best" when taken up within "a community"—for instance, how does any group of persons avoid 'group think' in these discussions; indeed whether any group can avoid that? Here, too, the argument simply takes for granted a full theory of social life and its articulation in medicine and especially clinical situations. And, it is just this that should on no account be passed over. For instance, whether "community" indicates a patient's church group, the patient's family (nuclear or wider), or some other social grouping, and so on will make a serious difference in decisions reached or even considered.

Second, Glover et al uncritically end up endorsing little more than merely another version of an 'applied ethics' model. They argue that the main role and goal of the ethicist is to "lead the discussion about the application of ethical principles in the particular case" (Glover et al. 1986, p. 24)—without, I must emphasize, in any way specifying just what 'application' itself entails or signifies. Just *whose* 'principles' these might be, furthermore, and which are 'appropriate' and why, how and according to which criteria such 'principles' might be selected, much less how any 'community' is or should be formed or how 'a community' of persons should go about resolving the inevitable internal conflicts—none of these significant issues are addressed, as they surely must be.

Third, viewing the notion of community from a slightly different perspective raises much the same problem urged against Jonsen and Toulmin: among the many voices—patient, family members, physician(s), nurse(s), etc.—whose viewpoint needs to be 'facilitated,' and why that and not others, much less what 'facilitate' actually amounts to in each clinical situation? That is, like so many in this field, privilege of position is simply given over to the physician. Clinical situations are viewed almost exclusively from the physician's point of view rather than that of others (patient, family member, nurse, etc.)—or, more accurately, from what the respective authors surmise must be the physician's viewpoint (since most of these authors are not themselves physicians). Just these shortcomings, I believe, are decisive reasons for rejecting any of these understanding of clinical ethics.

References

Agich, George J. 1990. A role theoretic look. *Social Science and Medicine* 30(4): 389–399.
Agich, George J. 2005. What kind of doing is clinical ethics? In Special issue on Richard M. Zaner, ed. Osborne P. Wiggins and John Z. Sadler. *Theoretical Medicine and Bioethics* 26(1): 7–24.
Aristotle. 1962. *Nicomachean Ethics.* Trans. M. Ostwald. Indianapolis: The Liberal Arts Press.
Bosk, Charles. 1985. The fieldworker as watcher and witness. *The Hastings Center Report* 15: 10–18.
Caplan, Arthur. 1982. Applying morality to advances in biomedicine: Can and should it be done? In *New knowledge in the biomedical sciences*, ed. W.B. Bondeson, H.T. Engelhardt, et al. Boston: D. Reidel.
Edwards, M.J., and S.W.J. Tolle. 1992. Disconnecting a ventilator at the request of a patient who knows he will then die: The doctor's anguish. *Annals of Internal Medicine* 117(3): 254–256.
Glover, J.J., et al. 1986. Teaching ethics on rounds: The ethicist as teacher, consultant, and decision-maker. *Theoretical Medicine* 7: 13–32.
Gorovitz, Samuel. 1982. *Doctors' dilemmas: Moral conflict and medical care.* New York: Macmillan.
Jonsen, Albert J. 1980. Can an ethicist be a consultant? In *Frontiers in medical ethics*, ed. V. Abernathy. Cambridge, MA: Ballinger.
Jonsen, Albert J. 1986. Casuistry and clinical ethics. *Theoretical Medicine* 7: 65–73.
Jonsen, Albert, and S.J. Toulmin. 1988. *The abuse of casuistry.* Berkeley/Los Angeles: University of California Press.
La Puma, John. 1987. Consultations in clinical ethics: Issues and questions in 27 cases. *Western Journal of Medicine* 146(5): 633–637.
La Puma, John, and E.R. Priest. 1992. Medical staff privileges for ethics consultants: An institutional model. *Quality Review Bulletin* 17–20.
La Puma, John, and S.E. Toulmin. 1989. Ethics consultants and ethics committees. *Archives of Internal Medicine* 149: 1109–1112.
La Puma, John, et al. 1988. An ethics consultation service in a teaching hospital. *Journal of the American Medical Association* 260(6): 808–881.
MacIntyre, Alasdair. 1981. *After virtue.* Notre Dame: Notre Dame University Press.
Orlowski, J.P., et al. 2006. *Journal of Medical Ethics* 32(9): 499–503.
Pellegrino, Edmund D. 1979. *Humanism and the physician.* Knoxville: University of Tennessee Press.
Schiedermayer, D.L., John La Puma, and S.H. Miles. 1989. Ethics consultations masking economic dilemmas in patient care. *Archives of Internal Medicine* 149: 1303–1305.
Schutz, Alfred, and Thomas Luckmann. I: 1973. *Structures of the life-world*, 2 vols. Evanston: Northwestern University Press.
Siegler, Mark. 1979. Clinical ethics and clinical medicine. *Archives of Internal Medicine* 139: 914–915.
Silber, Tomas J. 1981. Introduction: Bioethics and the pediatrician. *Pediatric Annals* 10: 13–14.
Toulmin, Stephen. 1982. How medicine saved the life of ethics. *Perspectives in Biology and Medicine* 25: 736–750.
Zaner, R.M. 1981. *The context of self: A phenomenological inquiry using medicine as a clue.* Athens: Ohio University Press.
Zaner, R.M. 1993a. Voices and time: The venture of clinical ethics. *The Journal of Medicine and Philosophy* 18(1): 9–31.
Zaner, R.M. 1993b. If you don't ask, you won't know. In *Troubled voices*, 119–136. Cleveland: The Pilgrim Press.
Zaner, R.M. 1993c. Accidents and time: The urge to be normal. In *Troubled voices*, 119–136. Cleveland: The Pilgrim Press.
Zaner, R.M. 1994. Phenomenology and the clinical event. In *Phenomenology of the cultural disciplines*, Contributions to phenomenology, vol. 16, ed. M. Daniel and L.E. Embree, 39–66. Dordrecht/Boston/London: Kluwer Academic Publishers.

Chapter 6
Voices and Time

6.1 Taking Time, Listening to the Voices

My critique of the casuist proposal serves to highlight several points of interest. With its insistence on practical reason as at once more grounded in our moral history and more capable of appreciating the exigencies of moral discourse, Jonsen and Toulmin's approach strikes me as very productive. It nevertheless has certain unanswered problems, as was suggested, and in the end remains too centered on the ethics consultant's relation to the physician—both of which are simply taken for granted as having priority as regards the identification and resolution of moral difficulties in clinical situations.[1] Hardly any attention is paid to those persons whose circumstances are most often at issue—patients and their significant others—and almost none to other clinical participants whose words and actions help to constitute the clinical encounter and clearly help to shape the moral issues embedded in any clinical encounter—nurses, physician consultants, therapists, and many others.

For all its problems, on the other hand, Glover et al.'s conception of the ethics consultant as decision facilitator working within "a community" seems an important ingredient to clinical ethics—although much remains to be systematically clarified about just what and whom that 'community' is or ought to include, whose voices should be given weight and why (as well as how this is to be done). Moreover, at every point, it is imperative that the ethicist be both accountable to and held accountable by those whose decisions are to be facilitated—they, after all, are

---

[1] It bears notice at the outset that many of the more than 2,500 consultations I conducted between 1981 and my retirement in 2002, involved serious cross-professional and cross-specialty issues: for instance, how quarrels with obstetrics affected which babies were after birth given over to neonatology for care; or, how care for critically ill patients was differently perceived by nurses and by doctors.

© Springer International Publishing Switzerland 2015
R.M. Zaner, *A Critical Examination of Ethics in Health Care and Biomedical Research*, International Library of Ethics, Law, and the New Medicine 60, DOI 10.1007/978-3-319-18332-9_6

the one who must then live with the aftermath of whatever decisions come to be made.

For all their obvious value, however, both these views (as well as the others considered in the previous chapter) conceive clinical ethics primarily from a single point of view: most often from what is taken for granted as the physician's viewpoint, but at times including the ethicist's perspective in relation to that of the physician. Both are problematic. First, neither emphasis appreciates the fact that every clinical encounter, even the simplest (say, a routine office visit), is essentially complex. As pointed out, each encounter includes multiple persons, each with their own set of concerns ('voices,' as I've termed them), and each of whom has (or claims to have) some 'say' in what occurs.

In part, the difficulty faced by the ethicist lies in the effort to explicate and assess that complexity and on that basis attempt to facilitate the conversation and eventual decision-making required by every clinical encounter. The ethicist faces a *specifiably complex* task: s/he needs not only to think and act in the most practical manner, but also must similarly help the other situational participants to think and reason in highly practical ways—about the clinical issues, options, decisions, aftermaths, and especially about what each of them takes to be *most worthwhile* within the constraints of the specific circumstances, as well as how to go about assessing that 'worth'.

Equally important, to reason practically means, among other things, to recognize the presence in any situation of the different, typical understandings at work in and emergent from the various participants' respective personal and professional experiences, interests, relationships to the special issues and to one another, etc. Hence, it seems clear, Jonsen and Toulmin's otherwise fine study must be critically deepened in order for it to be capable of illuminating the very clinical encounters that prompted its revival in the first place. Working out an appropriate "paradigm" to which any "case" might refer (and precisely how and why this or that "case" can and should be regarded as "appropriate" to the "paradigm," etc.) invokes a complex sense of practical reason and a fund of experience far richer than is recognized in their work. This is true especially if one invokes an ambiguous paradigm, or when a particular case involves a conflict among potential and competing paradigms. Indeed, it seems wise to resist a too-hasty move to the paradigmatic, for fear of missing precisely what is at issue: the search for analogies harbors more risks than they appreciate.

6.2   The Unique, the Similar

In fact, their emphasis on reasoning by analogy (as the basic sense of *phronesis* which, as I've already suggested in the second Chapter, ignores significant components of the mode of medical reasoning termed "*semiosis*" by the ancient empirics and later Skeptics), and this risks obscuring the most prominent characteristic of clinical encounters. They argue that clinical medicine provides "a powerful model" for understanding the ways in which practical and theoretical matters are related in ethics. In medicine, a description of a condition is clinical fruitful, they say, "only when it is based on perceptive study of actual cases, and it is practically effective only if paradigmatic cases exist to *show* in actual fact what can otherwise only be *stated*: namely, the actual onset, syndromes, and course typical of the condition." Diagnosis is then for them a type of pattern recognition, or "syndrome recognition: a capacity to *re*-identify, in fresh cases, a disability, disease, or injury one has encountered (or read about) in earlier instances," and the "reasons justifying a diagnosis rest on appeals to analogy" (Jonsen and Toulmin 1988, pp. 36, 40–41; Toulmin 1982).

This so-called "appeal" is not, I think, at all accurate; it is in any case surely inadequate, even risky. While not able to ignore such analogies and patterns, obviously, clinical practice is focused specifically on what is unique and individual: *this* unique individual now being diagnosed, cared for, and so on. Clinical practice is thus textured by a dialectical tension between the appeal to similarities (pattern recognition, analogy; more accurately, the *typical*) and the need to be specifically attentive and responsive to (as well as responsible for) each unique individual in his or her own unique circumstances. *This* unique person, after all, is who must be understood and helped, even while he or she, as well as the presenting illness or lesion, may show certain similarities to other persons.

This complex focus of clinical attention does not in the least belie the need to be capable of *showing* and not merely *stating* what's going on in the particular encounter, nor that this attentiveness requires and builds on a fund of experience that provides the physician with "paradigms." But it is precisely this dialectical play between past experience and now-presented individual patient which gives "pattern recognition" its strictest sense and helps to define clinical reasoning as preeminently practical, a clear instance of *phronesis*—or, to keep pertinent references within medical reasoning, a clear example of *semiosis*, which may well be understood as the more embracing concept. However that may be, as a therapeutic discipline, its focus is on the unique individual—who must never be forgotten in that tempting web of 'similars,' analogies, and paradigms, as ancient skeptical physicians knew well (Zaner 1992, 2001).

It is the same for clinical ethics: the individual encounter presented with its specific set of unique circumstances is the central and abiding focus of attention. It is to each specific situation and each individual whose situation it is that the ethicist must be responsive to and responsible for, even while this responsiveness and responsibility must surely be dialectically framed and informed by that fund of prior experiences—whose shadow, as it were, is invariably cast over every present event and thanks to which similar situations do indeed stand out as providing practical help for understanding and decision making.

6.3   An Interlude: On Wholes and Parts

It has been pointed out that every clinical encounter is highly complex in specifiable ways. Just because of the complexity, furthermore, the ethicist is faced with a specific task which requires that clinical reasoning be understood in an ever richer sense: namely, the need to be attentive to the situational encounter as a *whole* while in no way ignoring the unique individuals whose situation it is.

Consistent with the seminal work of Aron Gurwitsch (1964, 1966) an encounter is a whole in the precise sense that it is the system of multiple interrelationships among constituents (Zaner 1981). Constituents are not lost in the whole, nor is the whole reducible to its parts; rather, they stand in a dialectical tension with each other, and it is this tensioned union, the "*system* of multiple parts," that constitutes the whole (Zaner 1979). To enter a clinical situation is to be confronted with a whole, a "contexture" in Gurwitsch's precise sense, (Gurwitsch 1964, pp. 105–54) that is, the *set* of relationships among clinical units, rules, procedures, standards of practice, as well as people, etc. This complex set of multiple interrelationships constitutes the encounter as the unique situation it is, and just this is the necessary, practical focus of clinical ethics. Some further detail is helpful, especially to contrast my own view from that of Jonsen and Toulmin and the others considered here and earlier.

Four main points are necessary to understand the notion of the whole (which Gurwitsch worked out in common with his colleagues in Gestalt psychology: the idea of "form"), Gurwitsch proposed a terminology to aid in distinguishing his concerns from others (such as the Gestaltists). He thus introduced the notion of "contexture" as a principle of organization of the experienced world of perception. To grasp his meaning, four important concepts are needed: (a) functional significance, (b) functional weight, (c) Gestalt-coherence, and (d) good continuation.

(a) Each authentic whole is intrinsically articulated into parts or, preferably, "constituents," and thus reveals some degree of organized detail, by virtue of which it stands out from the field. I hear, say, a chord thanks to the fact that it stands out as a unity from a background of other sounds (for instance, a cough or sneeze). Specifically, a contexture exhibits constituents that have their systematic placement within the whole. To be a constituent (a part of a whole) thus *means* to occupy a certain locus or place defined strictly in reference to the topography of the whole.

This absorption or placement within the whole gives each constituent a specific *functional significance* for the contexture: for instance, 'being the right-hand member of a pair;' or 'being the third note in a minor chord.' Accordingly, "the functional significance of each constituent derives from the total structure of the Gestalt, and by virtue of its functional significance, each constituent contributes towards this total structure and organization" (Gurwitsch 1964, p. 116). Only as thus integrated along with other constituents into a whole, and as systematically related to the others that are also related to each other (according to the same principle) and to the first, is a 'part' a constituent of a contexture.

Should a constituent be removed (in whatever way[2]) from its contextural placement, situating it within another, one cannot speak "of the same constituent being integrated into different contextures" (Gurwitsch 1964, p. 121). For instance, if a C-major chord is heard, and then a C-minor one, the note 'G' constituent to the first is not heard as 'the same as' the note 'G' constituent to the second. Although, Gurwitsch admits, there is *a sense* in which 'the same' objective state of affairs obtains, this is not the case for auditory (or any other mode of sensory) experience. Since the latter is precisely the issue, it would be a grievous error to confuse the two. What is at issue, in other words, is the functional significance, and in the example given precisely this is what alters. "*It is the functional significance of any part of a Gestalt-contexture that makes this part that which it is*" (Gurwitsch 1964, p. 121).

(b) Consider the way a red stoplight is experienced when seen during an urgent drive to take your injured child to a hospital. Clearly, not every 'part' has the same significance within this contexture. The light has greater *functional weight* in this example than it does, say, when you are merely driving along in a leisurely manner. What is 'crucial,' as we say, has greater functional weight, and this is getting your child to medical help. It is in reference to concern, thus, that the light stands out as 'emphasized,' weighted—a veritable obstacle. Such weight is, of course, relative: that is, relative to the functional significances defining the other constituents. As he points out, "This import is in proportion to the contribution which, by virtue of its functional significance, a part makes to the contexture." (Gurwitsch 1964, p. 133), and is thus constituted in reference to the contributions of the other parts.

(c) It then becomes evident that the 'whole' or contexture is not the additive sum of its parts, nor is it reducible to its parts; nor for that matter is it 'more' than its parts. All such expressions are grievously ambiguous. A whole or contexture requires, in Gurwitsch's words,

> No unifying principle or agency over and above the parts or constituents which co-exist in the relationship of mutually demanding and supporting each other. The Gestalt...is the system, having internal unification of the functional significances of its constituents; it is the balanced and equilibrated belonging and functioning together of the parts, the functional tissue which the parts form...in which they exist in their interdependence and interdetermination. (Gurwitsch 1964, p. 139)

Every constituent not only refers to every other one, but also to the totality formed by that system of references; as any part is related to every other part, it is therefore also related to the *fact that* the other parts are similarly related. Hence, 'relation' here, is specifiably complex (Kierkegaard 1944/1957; Zaner 2012). A contexture or whole is precisely the system of mutually interdependent and cross-referential constituents or parts; it is this system of complex references or functional significances. Thus, not only does every part refer to every other part, but the whole is inherent to every constituent: precisely in virtue of its specific functional

---

[2] There are, as Husserl and, following him, Gurwitsch demonstrate, parts that cannot be thus separated from other parts: for instance, the color and extension of an object. Gurwitsch was critical of Husserl's way of attempting to account for this type of difference.

significance, each part 'realizes' and 'references' in its own specific way the whole contexture.

(d) Gurwitsch points out that it is the contexture (which he terms "theme") which makes possible the organization of the context (the "field") as materially relevant and as background for that theme. But what makes the theme itself possible? Several conditions have already been pointed out.

(i) Although the theme makes possible the organization of the field, it is reciprocally the case that every theme appears as within and standing out from its specific field. Thus, Gurwitsch points out that in the case of perception, "*per-cipere* may be characterized as *ex-cipere;*" (Gurwitsch 1964, p. 321) it is a "singling-out" of the theme from the field. In different terms, the "ground" can never be absent from perceptual "figure" (Gurwitsch 1964, p. 113).

(ii) The theme does not merge into, but emerges from, the field. *Not* to be absorbed into the field, thus, signifies the specific kind of "coherence" displayed by contextures—a 'being-bound-together,' as it were which does *not* hold among items in the field, or between the field and the theme. The segregation of themes from the field follows the lines of and "is a condition of" segregation" (Gurwitsch 1964, p. 138).

(iii) Every theme has a certain "positional index:" an orientation, position, or placement within the field. For instance, a particular proposition is (it has the functional significance of being) the conclusion of an argument. Its positional index consists of what Gurwitsch calls "contextual characters:" for example, "referring back" to premises as "derived from" them, and "referring forward" to other propositions, etc., all within the field of logical relationships among propositions. The theme appears within the field; it has a certain "position" within the field and thus serves to orient the field.

(iv) The field is thus not undifferentiated. Simply focusing on one thing (a house, a proposition, etc.) does not render the field of other items into an amorphous vagueness: consider, for example, the items in a room while you are focused on a particular painting on the wall. These other items in the field remain relatively distinct and definite, differentiated from still other items, even though not now attended to or thematized. In short, it is part of the organization of the field that each of its items is itself a potential theme—which is part of the meaning of material relevancy. When thematized, the item retains its sense of having been materially relevant, of having been potential. Briefly, then, the central conclusion follows: the organization of the field into theme/thematic field/margin is not 'derived' from anything else, but is rather, Gurwitsch says, autochthonous; (Gurwitsch 1964, pp. 30–36) it originates precisely *there*, where it is found.

Gestalt psychologists had already identified four factors that determine the organization of wholes. In ascending order of import these are: proximity, equality, closure, and good continuation. Although first established as regards only visual wholes, Gurwitsch shows that they have significance far beyond that. His analysis to this point already showed in effect that the first two (proximity and equality) are comprehended by functional significance and coherency; closure and good continuation remain to be accounted for by his proposed 'field' theory of consciousness.

These two can best be elucidated in cases of incomplete contextures: e.g., a melody broken off before completion, a sentence left dangling, a face incompletely drawn, etc.

In each case there is an experienced incompleteness and a pronounced tendency toward completion (closure). This tendency, however, occurs solely along lines already laid out by the presented, partial contexture (good continuation). All incomplete contextures appear as "in need of support and supplementation...in accordance with their functional significance" (Gurwitsch 1964, p. 151). The actually given constituents include what Gurwitsch had already termed "pointing references," but in the case of incompleteness, these references are toward other constituents as needing to be at certain places and with certain functional significances in reference to those at hand and thus in reference to the as yet incomplete contexture. Clearly, not just anything will serve to complete a melody (e.g., the noise of a passing train), or a sentence (e.g., the feel of a rough texture), or a drawn face (e.g., an odor in the room). The incomplete contexture "develops strong tendencies of its own toward completing itself" by setting out what sorts of constituents would or would not 'fit' into itself (Gurwitsch 1964, p. 151).

Such cases of incompleteness help make clear what even a well-formed contexture possesses but is not always easy to detect. Contextures have a striking tendency to persist and maintain themselves, and in this sense toward preserving their integral concord or coherence. Should such continuation fail to occur, thus, an incongruity is experienced, a being-out-of-tune, a clash and discord characteristic of abortive or flawed contextures—or, as might also be suggested, of impaired embodiments or mental life.

It is thanks to this tendency to good continuation and closure, that contextures present a kind of strength or connectedness, a remarkable unity. Yet, while each is thus a 'one,' a *unity*, each is also an intrinsic *diversity*, a 'many.' Systematically and functionally placed within the topography of the whole, each constituent is nonetheless differentiated from every other one. They also differ from the total system of functional references: each constituent is positioned in its own way, and each presents the entire contexture from its own position. Yet it is solely by virtue of the contexture that each constituent has its specific position, functional significance, and weight.

That is, diversity and unity are mutually conditioned and conditioning: a contexture is necessarily a unity-in-diversity, since it is the systematic significance of each constituent to be at once 'itself' and 'different,' and essentially to be a complex referencing and being-referenced vis-à-vis the total system (the whole). This could be seen, then, as Gurwitsch's response to that traditional conundrum, active since at least early Greek times, the problem of the 'one' and the 'many'.

6.4   Return to Clinical Ethics

Gurwitsch's field theory is by any reckoning a powerful instrument for understanding the genuinely complex phenomena of perceptual experience, including clinical experience. That it is equally powerful as regards other objective phenomena seems quite as clear. He argues only, as mentioned, that it addresses the organization of the noematic-objective sphere of sense perception—the phenomenological term for the 'what is experienced, precisely as and only as it is experienced—that of experienced objects in general, and thus in no way usurps the principle of temporality except as the argument is that the latter is now restricted to the noetic sphere.

But precisely in view of the impressive way in which his phenomenological theory is able to illuminate hitherto obscure or poorly understood phenomena, I have long been naturally led to wonder about its extension—i.e., to ask whether that restriction to the objective sphere of sensory perception is needed. This is neither idle speculation, nor an ad hoc way of engaging in cleverly sportive argumentation. In my work after studying with Gurwitsch, and studying his seminal writings, I have tried to do what he did with Husserl: to carry out the sense implicit to and consistent with the original.

For example, I have already suggested in some detail (Zaner 1981) that the notions of context and contexture are highly significant in that they usefully elucidate otherwise very puzzling phenomena relating to mental life, self and embodying embody. Nor would I be the first to find such extensions to such generative notions. After all, it could readily be pointed out that not only thinkers such as Wilhelm Dilthey, but Husserl and even Gurwitsch himself, frequently use descriptive locutions when referring to the sphere of self and consciousness which immediately suggest the very organizational principle used to articulate the noematic field. Thus in Dilthey's programmatic but intriguing essay on descriptive and analytic psychology, (Dilthey 1894/1977, pp. 20–21) and elsewhere in his work, clearly one of the basic concepts is the "nexus" of mental life (*psychische Zusammenhang*) and much of Dilthey's concrete descriptive analysis is strongly suggestive of the contextural principles Gurwitsch has delineated. Husserl, too, is often obliged to characterize consciousness as, in his term, a "concrete context of subjective mental life" (*konkreten wesenseinheitlichen Zusammenhang eines subjektiven Erlebens*) (Husserl 1928/1969, p. 157).

All of which states the matter somewhat abstractly. But what has been explicated above can be readily appreciated in any of the narratives that have been referenced in these Chapters. Indeed, every clinical encounter makes the point dramatically. Every such encounter invariably includes many situational participants ("constituents"): patients, families, and friends; also physicians, medical consultants, nurses, chaplains, social workers; even the clinical ethicist is obviously among the contextual constituents of the encounter. All of these persons have some legitimate (though perhaps not always clear) stake in analyses, decisions, outcomes, etc. and the variety of relationships among them frequently needs careful identification, sorting, and assessment. As gradually became clear to me, moreover, the 'language' most suited for expressing these highly complex and concrete interrelationships is that of *narrative*.[3]

Many of these issues, obviously, come into clear focus through the efforts to communicate the patient's diagnosis, possible therapies, likely outcomes, etc. It seems to me that the ethicist's main role is not what Agich terms the "watcher," nor is it as the analyst of analogies and similarities (Jonsen and Toulmin), nor facilitator (Glover et al). The clinical ethics consultant is, rather, the participant who is invited by one or another of those already positioned within an encounter, and whose task to identify and help others to understand the variety and complexity of discourses going on within the clinical encounter—as well as whatever else may be associated with any of those discourses: goals, histories, and the like. In addition, the clinical ethicist attempts to help the various participants to understand each other (for their own sake and that of others), and to focus on whatever moral issues can be shown to be ingredient to the encounter, as clearly and amply as the always constrained and constraining circumstances permit.

In other words, I can now say, the central focus of the clinical ethicist is the specific whole, the contexture of multiple interrelationships that make up the specific encounter. It is the ethicist's task, perhaps only ideally in many encounters, to enable or promote relationships that are most consonant with the respective participants' own moral and/or professional traditions and commitments.

Clinical ethics issues are presented, as I've urged elsewhere, (Zaner 1988/2002, pp. 242–50) strictly within the contexts of their actual occurrence, and these contexts are complex sets of relationships, principally among persons of various sorts, some being primary decisional stake holders, others less so, and some not at all. The focus of clinical ethics consultations is that whole, that contexture of relations—in the narrowest sense, that between physician and patient is also a contexture—the aim of which is to pursue, enable and promote quality patient care.

6.5 Facilitating Decisions

Primary decisions fall to patients (sometimes, to families and/or significant others), legal surrogates, physicians. The ethicist is not a decisional stake holder; his or her role is to facilitate the complex conversational process, specifically by helping decision-makers become aware of and to think about the clinically presented moral issues in the most profoundly practical manner: to understand and be understanding toward one another, and eventually to reach decisions from within their own respective moral frameworks (what is deemed as 'worthwhile'), with the aim of reaching decisions with aftermaths that are as consonant as possible with

---

[3] As gradually became clear to me, moreover, the 'language' most suited for expressing these highly complex and concrete interrelationships is that of *narrative*.

each participant's own respective moral framework and within each person's own particular circumstances.

Clinical ethics consultation is in some ways similar to teaching introductory philosophy. Like most beginning college students in an ethics class, most patients (and their families, at times their physicians and others) have rarely if ever gone through the process of thinking about their own basic sense of what is truly worthwhile for themselves. Among other things, the ethics professor must have the skill to help students develop the courage, skill and insight in order to learn to do what they have rarely (or never) done, realizing that the process will invariably be immensely challenging, even daunting. Similarly, the ethicist must have the sensitivity and skill to conduct this process with at times vastly different sorts of people—patients and families, physicians and nurses, etc.—in situations that can be critical and will often be emotionally charged.

Typically, of course, people for the most part rarely have to do this; they do not have to think deeply about such matters; even when such occasions arise, the thinking is not always done with great skill, to any great depth, nor for very long. Illness, injury, or handicap (whether from genetic, congenital or social circumstances), however, frequently requires precisely such serious, in-depth thinking about what is most worthwhile. These challenges are not only novel for many people, but can be unnerving: attention has to be directed to what is truly fundamental for the persons themselves, for they can be, and often are radically challenged by their circumstances. What a person believes is most worthwhile can be, and often is, called into question, with decisions not only required but at times needed quickly. Here, ethics consultation is obviously quite different from classroom teaching. Time is often of the essence, with much critical thinking and deliberating needed quickly, with persons without experience in these activities. What is at stake is also quite different, for here things matter very much—with pain, suffering, life, permanent loss, grief, even death hanging on decisions made or not made.

Nevertheless, the ethicist is asked to help not only patients and families but health professionals as well—to help them do what is for most of us neither habitual, native, nor easy: search into their own most basic sense of worth of self and other, always challenged by impairment which can itself defy the ability to reflect. The point is to attempt to find ways to identify, articulate and understand this core of what is held to be most worthwhile as fully and fairly as circumstances permit, as well as to locate options, decisions, and outcomes that are consonant with that core of value—as vital for any actions, such as deliberating which decisions should then be made.

Finally, although physicians[4] are themselves placed *within* the primary relationship with patients and families, their own matrix of moral belief is not itself the

---

[4] And, to some extent, nurses, consultants, and other health professionals, but especially those nurses who are involved in primary care for patients. This of course varies somewhat from hospital to hospital.

specific focus of concern, although it surely can be (in which case, there is a wholly different set of moral and other issues that will need to be faced and resolved). The patient and family, in contrast, are focused by the patient's medical problems—perhaps others as well, such as financial questions—and toward getting better, to whatever extent this is possible (and if not that, then on comfort, control, dignity, etc.). The physician and other health professionals are focused on understanding and taking care of the disease process and/or injury, and toward caring for (benefiting and not harming) the patient.

It is thus unrealistic at best to expect physicians or patients/families to be capable of conducting deliberative, probing conversations about the respective fundamental sense of worth each participant may hold, not to mention hopes, fears, or issues of trust. An essential component of these moral conversations is concerned not only with the patient and family, but with the physician's own professional responsibilities and personal ethics—even, at times, with institutional values and social norms.

Thus, La Puma's proposal that the ethicist must be an experienced physician is deeply problematic: acquiring knowledge and skills such as those mentioned, learning to focus systematically on the network of interrelationships among all decisional participants, understanding how to go about identifying and grasping fundamental moral commitments, and the like, clearly suggests that being a physician does not of itself qualify one as an ethicist; it may, in fact, interfere with the process. For that matter, it is dubious to suggest that tacking on a 1-year postgraduate fellowship in clinical ethics—whatever else may be gained from these well-known periods of study—will provide the necessary skills, intelligence, or background. The clinical ethics task, in a word, is *not* to bring some presupposed set of 'ethical principles' to clinical situations—there, somehow to be 'applied'—but rather to be capable of *discovering* and then giving voice to what already morally textures every clinical event. Every situation presents participants' own moral views and it is these that must be uncovered, articulated and understood in the prevailing clinical circumstances. La Puma's claim to know the necessary requirements for certification in clinical ethics thus strikes me as quite premature, if not presumptuous (La Puma and Toulmin 1992, p. 19).

Although there is much to say for the physician serving as the patient's advocate, it is nevertheless the case that the physician is essentially *a part of the relationship* with a focus on helping and not harming each patient. Both patient and family (or legal surrogates), and physicians, thus, are the primary decision makers, in whatever way the particulars of this are worked out in specific cases. Neither of them is as such focused on the complex of relationships itself, nor on how that affects the respective moral frameworks and decision-making.

What is at issue for the clinical ethicist, however, is different: neither patient, family, nor physicians are focused on the specific complex moral sense of each relationship (family/patient, patient/physician, family/physician, etc.). Just these, however, are the fundamental concern for consulting ethicist. If the patient is oriented toward getting better (or feeling less pain and suffering), and the physician toward the patient (helping, not harming), the ethicist's more complex orientation is that contextured set of relationships, whose understanding and resolution the ethicist seeks to facilitate precisely by sensitive, even at times overpowering conversations on and about the multiple senses of worth presented in the special circumstances of the clinical situation.

6.6   Clinical Conversation

Working within a resounding legion of voices, often impressively dissonant (though it can also at times be surprisingly congenial), is surely among the most striking factors about being involved as an ethicist in clinical encounters. Thus the image of 'voices'—challenging, compelling, urgently seeking to be heard—and the exigencies of time—time to think about matters, time to speak and be heard, time to listen, not enough time to settle disrupted things—are themes that run through these reflections, because they run through every clinical encounter.

Suffusing a patient's words—a father's grumble, a mother's lament, a baby's whimper, a grandfather's relinquishing sigh—is the voice of vulnerability, someone seeking and hoping for help and/or restoration, a voice marked by a compelling narrative: the telling of self imperiled in the face of loss, grief, pain, decline, ultimately death, while in the hands of people who are not sick, not in grief, not in pain, and do not face death in any immediate way.

From within the patient's circumstances, as I've emphasized, the relationship to her physician is indelibly marked by the *pathos of unavoidable trust* (Zaner 1988/2002, pp. 53–56, 69–71). By contrast and despite their technical character, the physician's words reveal the voice of ability, of knowledge and know-how, of power and control in the sense especially of a confident *ability to do* governed by beneficence and *nolo nocere*—more particularly, *taking care of* (hopefully even caring for) people who cannot take care of themselves. It is at times perhaps a strident confidence permeated by a sense of being professionally trustworthy, and yet sometimes also a doubt—even self-doubt, an unsureness masked by the guises of professional certainty and control. From this point of view (at least in its finer moments), the relationship to patients is marked by a sense of healing and caring, (Pellegrino 1983) and sense of being trustworthy: competent, sensitive, responsible for and responsive to patient needs and concerns.

Among the other voices choired within any encounter one can readily discern the reverberation of moral feelings embodied in images, noises, and gestures, expressed in personal and social discourses, and the urgencies to be heard, even merely noticed. The ethicist's work, in part at least, is committed to finding ways of enabling appropriate and timely hearings, oriented in and around the centering relationships of patients and physicians. If, as Eric Cassell once put it, "the spoken language is the most important tool in medicine" (Cassell 1985, p. 1)—and I think he is right—it is most important for the ethics consultant to cultivate sensitivity, and skillfully to orchestrate the multiple discourses ('voices') in clinical encounters (medical, biomedical, moral, religious, as well as everyday talk in the specific voices of patients and those within their circle of intimates), ensuring, so far as possible, the time to be given an appropriate hearing so as to promote the basic aim of quality patient care.

The physician's work is governed by being at the behest of vulnerable people; it is thus fundamentally governed by assiduously *respectful competence* and *care*. The ethicist is similarly at the service of the various people involved in multiple and complex ways, and is fundamentally governed by *compassion*. This orchestration of the multiple voices is most properly conceived, therefore, as the work of a special form of dialogue which, within the constraints and exigencies of clinical encounters, is conducted within shared narratives, each with their own moods and languages, as the case may be, governed by such virtues as compassion (affiliation) and courage—all of which is in need of much deeper and careful probing. It is thus necessary to understand what it means to be ill, what the illness experience is all about (Frank 1995; Kleinman 1988).

6.7   Illness and Vulnerability

Illness is remarkable and disturbing, for unlike almost anything else in our lives it uniquely *singles me out as this individual*, as who and what I am: the bodily pain, the anxiety over future prospects, the way in which my condition occupies (and preoccupies) so many people around me, the way so much of my daily life has to be reorganized and rescheduled, and the riveting focus of the pain and strange new feelings. How did *I* come to be *this* sick person, embodied, embedded even, in this body which hurts and can no longer do what I need to have done: how did *I* get here? (Zaner 1985)

In some ways, this singular experience has its source in the sheer *happenstance* of illness: it befalls me without my having wanted or chosen it. To try to find some reason for my having fallen ill, in the end, is to try to find some reason for being myself. In some sense, too, my becoming ill has its source in what is forcefully demonstrated by it: not only is each of us affected in basic features of our humanity, even more fundamentally it marks each of us as *able to be lost,* ultimately as *able to die*, as threatened by and exposed to *not being*. In everything I am and was and ever hoped to be, I find myself exposed to my own not-being, to the finality of my own life. This is a powerful and unique singling out of the individual as the person he or she is through telling glimpses of loss and death, of no longer being that person: hence, the illness experience is marked by the disclosure of the person's own intimate *vulnerability*, ultimately to no longer being who and what I am.

To experience oneself as vulnerable—and in part to undergo the experience of needing to know what's happened and, too, to want to be cared for—is thus to experience oneself within a kind of appeal that seeks response. To be vulnerable is to want to be known, to seek recognition and help: *I* am sick and *I* need help. Not only this, however, for experiencing illness, I need help from the other, from *you*: in my vulnerable needing of help, I now turn to the other. I who have been singled out by illness now single you out in my appeal for help. To want care, in this deeply personal sense, is *to want to be this self* in *your* eyes and hands—in the eyes and hands of this nurse, this doctor—just as trust is a deeply personal wanting to know and be with this nurse and this physician as the persons they are. Illness in this sense bodes the promise of a special sort of 'us,' a 'we' who now together seek ways to respond to this appeal that is me.

Illness is uniquely *complex*; it is an appeal for responsive recognition and help by one person from other persons who are looked to for help. This therapeutic dyad, caring and trusting, is thus a profoundly intimate, moral phenomenon, whose ground seems most fundamentally the vulnerability that textures every human life but is especially marked out by affliction. The promise of this relationship is that not only may the sick person recover from the illness; more, it promises the *recovery of ourselves,* patients and caregivers, *as persons* who are 'worthwhile' and cared for. The promise may not be fulfilled; it may be broken, ignored, violated in many ways, or not even appreciated for what it is (and this for a variety of reasons). Still, this promise of the recovery of selves is ingredient to the illness experience: to threat of loss, to grief, even to death and dying.

While the relationship between physician and patient is rich with fiduciary promise—which, admittedly, is fundamental to professional integrity—it does not in the first place evoke the primacy of the physician. Rather, the integrity of the clinical encounter, if you will, arises as a response to the initiating act of trust by the patient. In this sense, the phenomenon of vulnerability harbors the clue to professional integrity.

This approach picks up on an insight in Gabriel Marcel's many writings, that such interrelationships are best understood as dialogical, in turn grounded in appeal and response, a key component of which is what he frequently analyzed as "availability" (*disponibilité*) (Marcel 1940, pp. 188–89, 1949, pp. 54–55, 1935, pp. 160, 180–81). Although these are closely interrelated in any dialogue, the appeal (for recognition, help, information, etc.) initiates the dialogical event by specifically appealing for or inviting a response. In this sense, to appeal is to request a response to (in the complex sense of a reason for, a story about, even, it may be, an evocation of) the patient's (the appealer's) condition, whatever it might be, and thus figures prominently in the integrity of the clinical encounter.

6.8   The Patient's Centering Place

The patient has a peculiarly commanding presence that contrasts strikingly with the structural imbalance of power inherent to the relationship with the physician. As already noted, the relation is asymmetrical with power (in the form of the ability-to-do, to effect change in the patient's condition) on the side of the physician, not the patient. The physician has the knowledge and skills necessary for treating; not the patient. The physician, not the patient, has access to resources (diagnostic technologies, surgeons, prescriptions, hospitals, etc.). Unlike patients, finally, physicians are legally authorized and socially legitimated to use their knowledge, skills, and access. Not only is the patient compromised by whatever condition it may be (illness, injury, distress, handicap), therefore, but is also disadvantaged by that very relationship.

Doubtless many relationships, especially those involving the helping professions, are asymmetrical—parent/child, lawyer/client, teacher/student, police/citizen, etc. When these are institutionally organized, furthermore, special issues crop up concerning the characteristic expectations governing activities within the institutional structure as well as relationships with the people outside the profession who seek help.

Still, there is something unique about the illness experience and the fiduciary relationship appealed for and (hopefully, but not always) invoked by it, as there is about the profession of medicine. Illness is an intrusive and capricious irruption in the individual's life, and the illness experience is marked by a sense of urgency that is underlain by the always-present threat of compromise and loss, ultimately of death. "The ill person," Pellegrino observes, "becomes *homo patiens*—a person *bearing* a burden of distress, pain, or anxiety; a person set apart; a person wounded in specific ways" (Pellegrino 1982, p. 158). Illness cuts into the fabric of the person's ongoing life, abruptly alters the usual relations with others, and compromises the person's sense of self and world.

The sick person must contend with these crises while on the other hand she is more or less vulnerable, at a time when she often may not know what's wrong or what can and should be done about it, and when her energies are focused by trauma, pain, shock, suffering, and distress. The sick person most likely doesn't even know (except, it may be, in the generalized sense of daily life's typifications) whether the one who professes the ability to help, heal, or cure can in fact do any of these. Simple possession of a license, degree, or other marker of competence does not of itself guarantee the ability to handle the problems specific to her condition, much less to be sensitive and caring in doing so. Hence, the patient's at times acute sense of dismay over whether she can actually trust this individual doctor, those nurses, or the technicians called on to perform certain tests, etc.—in the face of which she may try to ignore her illness and pain, or to rely on her own resources for help. Except for that, however, help from another person must be sought, and she thereby enters into that structurally asymmetrical relationship.

In every society and historical era, there is a critical common factor in the encounter between professed healers and patients: the need for healing. "Medical thought grows out of, and is governed by, therapeutic experience. Therapeutic theories in all their variety are attempts to makes sense of the healer's experience with the patient" (Coulter I 1973, p. viii). In different terms, to think about the clinical event is to discover "the universal fact that humans become ill and in that state seek and need help, healing and cure" (Pellegrino 1982, pp. 157–66). It is worthwhile to cite again a central passage in Pellegrino's work:

> The healer professes to possess precisely what the patient lacks—the knowledge and power to heal. The healing relationship is thus inherently one of inequality which the patient enters in a special state of vulnerability and wounded humanity not shared by other states of human deprivation and vulnerability.

To be sure, the poor, the imprisoned, the lonely, and the rejected are also deprived of the full expression of their humanity, so much so, that men in these conditions may long for death to liberate them. But none save saints seeks illness as the road to liberation. In no other deprivation is the dissolution of the person so intimate that it impairs the capacity to deal with all other deprivations (Pellegrino 1982, p. 159).

Before probing these matters more deeply, it is important to gain firm purchase on what an ethics genuinely responsive to issues that occur within clinical encounters in the sense indicated must centrally incorporate. This may be expressed in the form of certain fundamental theses, which I had earlier expressed somewhat differently.

6.9   Theses of Clinical Ethics

Thesis 1: The work of ethics requires strict focus on the specific situational understanding of each involved person.

What a particular situation is—which values and what weight are attached to the components of the situation (objects, people, relationships, etc.)—is strictly a function of the experiences and interpretations of those whose situation it primarily is. This is, of course, a modification of W. I. Thomas' classic thesis about social life: "If men define situations as real, they are real in their consequences" (Thomas 1928, p. 572). This requires what Alfred Schutz, following Max Weber, called the "subjective interpretation of meaning" (Schutz and Luckmann 1973, pp. 243–99). If one wants to understand a situation, there is nothing for it but to try one's best to get at the ways in which the situational participants themselves understand their situation, and endow its various components with sense and meaning ("functional significance" and "weight," as was noted earlier). A highly significant modification of this classic thesis will have to be introduced sooner rather than later: so far as I am involved in such a situation, my own sense of it—initially and throughout—is as well an intrinsic component, although the sense of this must still be worked out with care and clarity. To say this in a brief phrase: every involvement by an ethicist in a clinical situation, whatever else it may be, is essentially *reflexive*.

Of course, to be involved in or be an actual party to a situation is one thing; to be an observer (as the philosopher-ethicist may be from time to time) is quite another. The problems presented by a situation are the problems of those whose situation it is most directly and immediately, just as are the alternatives, decisions, and aftermaths. Just as the physician is charged with acting on behalf or in the interests of each specific patient (and, often, family) within his or her own specific set of circumstances, so is the work of ethics in clinical situations under the requirement of acting on behalf of the situational participants—whatever can at some point be said to have initiated the involvement. That set of persons includes not only the patient, family and circle of intimates, but also the physician or physicians (interns, residents, consultants, etc.), nurses, and other care providers, as well as (where appropriate) the hospital and its units, always within the broader context of prevailing legal and governmental policies and prevailing social norms. Whatever contributes to the particular situational definition of each relevant participant must be identified, considered, and weighed.

Certain of these contributing components will obviously vary from situation to situation, and these differences (however slight they may be and from whichever point of view) are critical for the work of ethics. While certain of these factors may well be vital medically and morally, not even those that appear only slightly important can be ignored. The latter may either be far more significant than appears at first sight; or, they may change in the course of a situation's temporal evolution. Furthermore, precisely because each situational participant experiences and interprets everything, to one degree or another (gestures, words, objects, pauses, objects, relationships, etc.) from his or her own particular perspective (or what Schutz terms his or her "autobiographical situation") (Schutz and Luckmann 1973, pp. 92–119) what is viewed as a problem, as alternatives for decision making, as decisive points in a situation, and the like will also vary. These varying situational definitions or understandings themselves require cautious and sensitive notice and oftentimes delicate handling in order to arrive at even a modicum of basic agreement.

Often treated as communication problems or breakdowns, such variable situational determinants very often involve such problems, but much more besides. Communication difficulties will often harbor more serious, deeper conflicts of interpretation, values, religious understanding, of life-style more generally. Problems of whatever sort arising from or embedded in the talk among participants may well be signals indicating those other, deeper-lying issues. The disciplinary work of ethics must thus be one of constant alert for just these matters—very often presented most subtly.

Thesis 2: Moral issues, at least those that must be taken up, are presented solely within the contexts of their actual occurrence.

In a word, one cannot expect to know in advance of a clinical encounter what will be expected, what may be demanded, what must be reckoned with and taken into account—even if one may, as a function of relevant past experiences, have a general idea of these demands and considerations.

In his two-volume study of patient and physician talk, Eric Cassell emphasizes that physicians (I would add: and other clinicians as well, such as the ethics consultant) must learn to be effective listeners as well as speakers. Studying how talk actually works in such everyday conversations, he lays out several of the tacit dimensions of such talk—such as its paralinguistic features, for instance, pause, speech rate, pitch, intonation, word choice, etc.—and notes the ways in which what is spoken reveals its own coherence and logic (Cassell 1985, p. 1). To understand what a patient tells you (including his or her intent and credibility), thus, requires being attentive to those paralinguistic features as well as to the actual words used. Accordingly, Cassell rightly insists that the clinician can and must cultivate the skills of listening as well as those of talking.

The central point of his study is that the clinician who learns to listen to a patient's language and paralanguage—to their ongoing conversation, distinguishing between what is heard and what that is interpreted to mean—is far better equipped to find out just what is wrong with each patient and to be more assured that the patient has been understood. Word choice, silences, incomplete sentencing, and the like are critical since they will invariably "change the meaning of words" used and convey a crucial part of the patient's intent—even when that intent is only barely expressed, or expressed only indirectly.

Nor is this by any means all that must be noted, for in talking about things, including aches and pains, hopes and fears, every speaker at the same time reveals him/herself—talking about these rather than those matters, for instance, at the same time expresses his or her definition of the situation, its meaning. Cassell remarks: "Learning to listen skillfully to the patient and to interpret judiciously what is said can be as critical as a diagnostic tool as learning to hear and interpret heart sounds" (Cassell 1985, p. 45).

Precisely the same is true for understanding the many kinds of moral issue, conflicts, dilemmas and problems occasioned by clinical situations. Indeed, to focus on the specific situational understandings or definitions that constitute the clinical encounter is in many ways to focus precisely on the phenomena of talking and listening among those involved in the situation (Zaner 1996). Moreover, apprehending what one or another participant says at a particular point requires understanding as well what the actual setting and its distinctive features contribute: whether it be the Emergency Room, the Operating Room, an office, a waiting room, their own living room, a bar; the manners of dress, the general appearance of the people, the objects and instruments and their arrangement, as well as the types of actions being performed or suggested, and what occasioned the specific encounter. Patients, physicians, nurses, and others interact differently in different situations, with different people (whether actually different or only treated as different). Participants also engage in such encounters with different interests, histories, and aims.

In short, despite the presence of numerous typifications, there are significant differences that must also be noted. To understand what's going on in a specific situation—what's troubling which people and at which stage of which encounter, what's on different participant's minds, etc.—requires alert, cautious probing of the multiple facets of each situation: the discourse and its paralinguistic features, the setting, the particular concerns and goals each manifests in countless ways, etc. Clinical understanding, whether for ethical or medical matters, calls for disciplined listening, including these paralinguistic, conversational, and contextual probings and assessments. The moral issues that must be noticed and addressed are presented solely within the contexts of their actual occurrence and therefore requires skillful and sensitive identification, probing, and attention to the full complexity of each context.

Thesis 3: Each particular situation is in its own way imprecise and uncertain, and the different types and dimensions of imprecision and uncertainty are critical for everyone involved.

Consider only some of these dimensions:

(a) There are types and degrees of *uncertainty and ambiguity* in every clinical situation—for instance, of diagnostic tests and their outcomes and the ways in which these are expressed, of potential therapeutic regimens, aftermaths of one or another type of treatment, etc. The same is true of expressions of desire, consent, preference, compliance, refusal, and so on. Moral discernment (so-called "practical wisdom"), on whosoever part it may be and precisely like clinical-medical assessment, requires the deliberate effort to gain greater clarity and precision about these matters, at least so far as circumstances allow, and at all times signals the need to make the fewest irreversible decisions and/or assessments possible at any given time.

(b) Each situation, furthermore, presents *multiple issues* framed within various emotional and volitional factors and calling for decision and action. This is so not only during the ongoing course of any clinical situation but also simultaneously. People only rarely are 'single-minded,' especially when there are numerous aspects, complexities and alternatives presented by each situation, each of which harbors various uncertainties. Like clinical-medical assessment, thus, moral discernment requires continual alertness to conditions and circumstances (of various sorts), which change in various ways partly due to determinations and decisions being continually made during its course.

(c) Every clinical situation is accordingly inherently subject to the *fallibility* of the persons involved and their various abilities and skills at discernment, reflection, assessment, decision, and the like (Gorovitz and MacIntrye 1976). Not only do people display differences in these always necessary skills, they also differ about goals, motives, and the rest of the distinctive human repertoire.

(d) Finally, each specific clinical situation is highly *complex*, a characteristic which leads to still other forms of uncertainty and ambiguity, since people will again differ in how a complex of factors should be sorted out, how priorities should be organized, and the like. There are still other dimensions to this complexity, as will shortly become clearer.

Thesis 4: Medicine and health care more broadly are governed by the effort to make sense of the healer's experiences with the patient, whose own experiences and interpretations are ingredient to what the healer seeks to understand and, eventually, treat.

As noted above, the medical historian, H. L. Coulter, has termed this characteristic the "therapeutic experience" which he contends governs all forms of health care and the physician's efforts in particular. In other words, the clinical situation is, as I've argued in other places, is at the very least complexly *dyadic*: it includes not only the encounter with a distressed or damaged person (patient) along with the effort to interpret presenting 'symptoms' so as to lay out one or more alternatives for the course of therapy. It also includes the patient's own experiences and interpretations of 'what's wrong with me,' along with a complex nest of values, motives, hopes, desires, goals, and so on. The would-be healer, in other words, tries to help: to heal, restore, ease pain, provide comfort, and the like–itself complex in its own way. The person-to-be-helped (patient) also responds to these efforts by the healer, and these responses themselves become a necessary part of what the healer must then "treat." Edmund Pellegrino expresses the point more simply:

> The end of the medical encounter, and the process through which it is achieved ... is restoration and healing—some corrective, remedial or preventive action is directed at what the doctor and the patient perceive as diminution of the patient's wholeness, each in his/her own fashion. (Pellegrino 1979)

Making sense of what the physician encounters clinically eventually finds its way into a clinical judgment, which is the result of responding to three large questions posed by each patient's presenting condition. First, what's gone wrong?, a question which leads to the diagnostic characteristic of clinical judgment. Second, what can be done about it?, which points to some range of possible treatments which can be brought to bear on what's gone wrong — the question of possible therapies. Third, what should be done about it?, which requires the joint efforts of patient and physician to reach a mutually agreeable decision regarding which of the possible therapies should be pursued in *this* specific instance. Pellegrino terms this the question of prudence.

I must add to his analysis, for it has always seemed to me somewhat elliptical: for instance, the 'what's gone wrong?' issue conceals several quite significant questions: what is it that brings this patient in to see this physician at this just time in the person's experience? Not only, if you will, has this patient come to the right physician for what the patient believes has 'gone wrong,' but equally pertinent is whether this physician understands the problems presented as appropriate to his or her area of competence. More about this later.

Thesis 5: This therapeutic theme is constitutive of every clinical situation, and embodies a fundamental moral resolve.

That resolve is enacted at the outset of the decision to become a physician or other health care provider, and enacted in every subsequent clinical encounter. It is, in a word, the resolve to put one's knowledge, experience, time and skills at the disposal of damaged or distressed human beings, individually or as groups or populations (hence the significance of Marcel's notion of *disponibilité*). It is a resolve, furthermore, revealing that knowledge and practice are uniquely integrated in every clinical situation and in medicine and health care more broadly. Medicine, in a word, is at one and the same time epistemic and therapeutic, every medical action is a matter of theory and practice simultaneously. More on this in what follows.

Accordingly, this resolve is fundamentally moral in character: it is another name for caring, in the sense of the resolve always to act on behalf of (and never cause harm to) a patient's best interests, the specific delineation of which in every instance is itself an inherent component of that resolve. It implies, furthermore, any of a number of possible *specifications* (or: specific responsibilities), depending on the demands of each specific clinical situation. For instance, if a physician is faced with a situation in which his or her limitations (knowledge, experience, background, technical skill) make the best interests or best course of action uncertain, then he or she has "an obligation to augment his or her knowledge so that benefits and risks of a particular regimen are as predictable as possible" (Liddle 1967). Similarly, the admission of limitations may also obligate the physician to seek consultation with others who have the needed sort of competence.

Of course, as noted, this therapeutic theme and its governing moral resolve on the part of the physician or professed healer is but one facet of the dyadic relationship.

Thesis 6: The existential condition of the patient is an inseparable (though distinguishable) facet of the dyad.

I have long emphasized that the dyadic relationship between patient and helpers is in essence asymmetrical with power on the side of the latter, especially in the case of physicians. I have already analyzed this characteristic imbalance in earlier Chapters, but it needs still more attention.

Consider first that only little reflection is needed to realize that this relationship is quite special. There is, for instance, Eric Cassell's observation:

> I remember a patient, lying undressed on the examining table, who said quizzically, "Why am I letting you touch me?" It is a very reasonable question. She was a patient new to me, a stranger, and fifteen minutes after our meeting. I was poking at her breasts! Similarly, I have access to the homes and darkest secrets of people who are virtual strangers. In other words, the usual boundaries of a person, both physical and emotional, are crossed with impunity by physicians. (Cassell 1985, p. 119)

Or consider Edmund Pellegrino and David Thomasma's emphasis that the ground for an ethics of trust (I only note here my critique of this and emphasis on "distrust")—which they regard as the heart of the "clinical event"—is the patient's "existential vulnerability." Although primarily focused on the patient's threatened organic systems (disease, injury, handicap), that very vulnerability obligates the physician to be keenly cognizant of and to respect the patient's moral agency—an obligation that increases as the invasiveness of medical procedures increases. Thus, when they contend, "it is the body of the patient that grounds this obligation, not merely social and legal structures," I think their point is elliptical, since it cannot be ignored that this "body" is the specific embodiment of this or that specific person. While a patient's "body is probed and violated in closer proximity and more intimately than is usually permitted even to those the patient loves," (Pellegrino and Thomasma 1981, p. 185) it is that patient herself, the one embodied by that sick or injured body, who is "probed and violated."

In any event, as noted many times already, one prominent characteristic of the clinical encounter is this *asymmetry*. Although the relation is in a sense reciprocal, it is nonetheless unbalanced with power on the side of the physician. Patients are constituted as *existentially vulnerable* not only by virtue of their bodily conditions, but also by the relationship's very asymmetry.

As was seen in an earlier Chapter, on the other hand, the relationship invariably involves profound *intimacies* regarding the body, the self, the family, their particular lives, beliefs, and circumstances. Since the relationship is most often among strangers, furthermore, the asymmetry and intimacy of contact can be especially tense, trust often problematic, and treatment open to compromise.

These themes set the context for a renewed explication of clinical medicine and clinical ethics and the encounters to which both must be responsive.

References

Cassell, Eric. 1985. *Talking with patients, vol. 1: The theory of doctor-patient communication.* Cambridge, MA: The MIT Press.

Coulter, Harrison B. I: 1973. *The divided legacy: A history of the schism in medical thought,* 3 vols. Washington, DC: Wehawken Book Co.

Dilthey, Wilhelm. 1894/1977. Ideas concerning a descriptive and analytic psychology. In *Descriptive Psychology and Historical Understanding*, ed Wilhelm Dilthey. Trans. R.M. Zaner and K.H. Heiges. The Hague: Martinus Nijhoff.

Frank, Arthur W. 1995. *The wounded storyteller: Body, illness, and ethics.* Chicago: University of Chicago Press.

Gorovitz, Samuel, and Alasdair MacIntrye. 1976. Toward a theory of medical fallibility. *The Journal of Medicine and Philosophy* I(1): 51–71.

Gurwitsch, Aron. 1964. *Field of consciousness.* Pittsburgh: Duquesne University Press.

Gurwitsch, Aron. 1966. Phenomenology of thematics and of the pure ego: Studies of the relation between gestalt theory and phenomenology. In *Studies in phenomenology and psychology*, ed. Aron Gurwitsch. Evanston: Northwestern University Press.

Husserl, E. 1928/1969. *Formal and Transcendental Logic.* Trans. Dorion Cairns. The Hague: Martinus Nijhoff.

Jonsen, Albert, and S.J. Toulmin. 1988. *The abuse of casuistry.* Berkeley/Los Angeles: University of California Press.

Kierkegaard, S. 1944/1957. *The concept of dread.* Princeton: Princeton University Press.

Kleinman, Arthur. 1988. *The illness experience.* New York: Basic Books.

La Puma, John, and S.E. Toulmin. 1992. Ethics consultants and ethics committees. *Archives of Internal Medicine* 149(5):1109–1112.

Liddle, Grant. 1967. "The mores of clinical investigation." Presidential address. *Journal of Clinical Investigation* 46(7): 1028–1030.

Marcel, Gabriel. 1935. *Être et avoir.* Paris: Fernand Aubier, Éditions Montaigne.

Marcel, Gabriel. 1940. *Du Refus à l'invocation.* Paris: Librairie Gallimard.

Marcel, Gabriel. 1949. *Position et approches concrètes du mystère ontologique.* Paris: J. Vrin, Philosophes Contemporains.

Pellegrino, Edmund D. 1979. The anatomy of clinical judgments: Some notes on right reason and right action. In *Clinical judgment: A critical appraisal*, ed. H.T. Engelhardt Jr. et al., 169–194. Boston: D. Reidel Publishing Co.

Pellegrino, Edmund. 1982. Being ill and being healed. In *The humanity of the ill: Phenomenological perspectives*, ed. V. Kestenbaum. Knoxville: University of Tennessee Press.

Pellegrino, Edmund D. 1983. The healing relationship: The architectonics of clinical medicine. In *The moral fabric of the patient-physician relationship*, ed. E.A. Shelp, 153–172. Dordrecht: D. Reidel Publishing Co.

Pellegrino, Edmund D., and David C. Thomasma. 1981. *A philosophical basis of medical practice.* New York/Oxford: Oxford University Press.

Schutz, Alfred, and Thomas Luckmann. 1973. *The Structure of the Life-World*, Vol. I. Trans. R.M. Zaner and H.T. Engelhardt, Jr. Evanston: Northwestern University Press.

Thomas, W.I. 1928. *The child in America: Behavior problems and programs*. New York: Alfred Knopf.

Toulmin, S.J. 1982. How medicine saved the life of ethics. *Perspectives in Biology and Medicine* 25(4, Summer): 736–750.

Zaner, R.M. 1979. The field-theory of experiential organization: A critical appreciation of Aron Gurwitsch. *The British Journal for Phenomenology* 10(3): 141–152.

Zaner, R.M. 1981. *The context of self*. Athens: Ohio University Press.

Zaner, R.M. 1985. How the hell did I get here? Reflections on being a patient. In *Caring, curing, coping*, ed. A. Bishop and J. Scudder, 80–105. Birmingham: University of Alabama Press.

Zaner, R.M. 1988/2002. *Ethics and the clinical encounter*. Englewood Cliffs: Prentice-Hall, Inc. Republished by Lima: Academic Renewal Press.

Zaner, R.M. 1992. Parted bodies, departed souls: The body in ancient medicine and anatomy. In *The body in medical thought and practice*, ed. D. Leder, 101–122. Dordrecht/Boston: Kluwer Academic Publishers.

Zaner, R.M. 1996. Listening or telling? Thoughts on responsibility in clinical ethics consultation. *Theoretical Medicine* 17(3): 255–277.

Zaner, R.M. 2001. Thinking about medicine. In *Handbook of phenomenology and medicine*, ed. A. Kay Toombs, 127–144. Dordrecht/Boston/London: Kluwer Academic Publishers.

Zaner, R.M. 2012. *At play in the field of possibles: An essay on free-phantasy method and the foundation of self*. Bucharest: Zeta Books.

Chapter 7
Responsibility in Clinical Ethics Consultation

7.1 Review

The field of bio- or medical ethics continues to show serious distress, appearances to the contrary notwithstanding. Not that such news will, I dare say, cause many tremors in the landscape of medicine or health care. What with the upheavals brought on by the open shift to commercializing health care—and the threats to the autonomy long enjoyed by many health professionals, physicians in particular, now experiencing the real anguish implicit to functioning as gatekeepers (Pellegrino 1986)—there has been quite enough to preoccupy physicians, researchers and others in health care's increasingly embattled demesne.

Turmoil among the practitioners of ethics consultation will not, I think, create much disturbance among health professionals or insurers. Ethics must seem of slight concern in the face of the harsh economic, social, and political issues in our times, especially those posed by the shift to so-called 'Obama-care', much less recent changes in medicare. Fierce competition for scarce dollars, resources and patients often define the continuing scramble by managers seeking to set up alliances among physician practices, HMOs, hospitals and so on. Health care has surely become the industry it has long been heading toward. The grander scheme of things has been scarcely moved by the internal (and, at times, even overt) disputes that have preoccupied so much of the literature in ethics—especially in what's come to be called clinical ethics. On such a massive landscape as that of health care in the United States, or other countries, images of small ponds and struggles over whose duck will squat and where, come readily to mind, on darker days, anyway.

However that may be, even more severe disputes than those I noted in the last chapters continue, some with considerable bombast. Still, although the scene may have changed it's not hard to detect several recurrent themes. Some are explicit—in particular, the vexing question whether a person involved in clinical ethics is a sort of 'moral expert.' Other questions, always nagging at the edges of one's conscience,

© Springer International Publishing Switzerland 2015
R.M. Zaner, *A Critical Examination of Ethics in Health Care and Biomedical Research*, International Library of Ethics, Law, and the New Medicine 60, DOI 10.1007/978-3-319-18332-9_7

remain a constant undercurrent, evident more by glancing blows of anxiety than direct address—I think especially of whether it makes any sense to hold an ethicist accountable for whatever is 'done', whether that 'involvement' carries with it any sense of responsibility.

Despite the fact that the quarrels over ethics in health care may seem only a somewhat pointless and in any case minor irritation to many health professionals, these questions nevertheless bear on that changing scene and themes of the new 'system' of health care. Although our moral compass may need serious re-calibration, if not wholesale overhaul, it has come to seem central to chart with great care such exotic new terrain as is plotted by those involved in the new genetics, especially that suggested by the key notion of responsibility in clinical ethics, not to say clinical medical practice.

In this Chapter, I wish to explore at least some of these matters, in particular a specific aspect of the more general issue that has been a recurrent if somewhat shadowy theme that continues to haunt a persistent dispute in the involvement of ethicists in health care, especially in that involvement known as clinical ethics consultation. As I am concerned specifically about the place and responsibility of ethics consultants, not so much as they serve on this or that hospital committee, but in particular within *clinical* situations, the questions of moral responsibility and legal accountability should be addressed strictly within the context of clinical conversation—which, it has long seemed to me, is best approached as a form of dialogue, (Zaner 1990) or, more recently, narrative (Charon 2006; Zaner 1994, 2004, 2012).

7.2 Who Is Responsible?

In previous Chapters, I went through much of the relevant historical background for clinical ethics. There, I noted that already with the first expression of what was at the time called medical (or: biomedical) ethics, there was a strong negative reaction by both physicians and many in the humanities. Struggling to contend with wholly new questions, physicians asked for help from philosophers and theologians in the effort, in the lingo of the times used by Dr. Samuel Martin, to 'humanize' practicing physicians (Martin 1972). Dr. Martin posed the essential questions—which, unhappily, have still to be seriously appreciated: what are the humanities really all about? How can they most effectively transmit their art? Widely regarded at the time as 'experts in human values,' a new, sibilant name was eventually concocted: 'ethicist.' But no more than named than they came under severe fire, and philosophers who dared to enter the new arena found themselves widely regarded as interlopers, merely theorists in the land of highly practical therapists, and were "well advised," as one philosopher admitted, "to limit their role as classroom or clinic casuists." (Ruddick 1981). They were not seen as especially helpful in the eyes of many physicians (Fleischman 1981).

At about the same time, I also noted, a widespread 'backlash' developed against what the former editor of *The New England Journal of Medicine* editorialized as "the intrusion of Big Ethics" (Ingelfinger 1975). As Daniel Callahan put it in 1975, there was a "sense that much of what is labeled 'ethics' represents a casual and irresponsible mischief-making, led by people with little understanding or research or practice" (Callahan 1975, p. 18). The point was not lost on many physicians.

It was, in fact, emphatically endorsed by Mark Siegler (1979) and turned with a vengeance onto medical humanists. As I noted earlier, arguing that they exhibit a clear "disdain for traditional, Hippocratic, bedside medical ethics," Dr. Siegler insisted that philosophers can never be more than mere observers at clinical events. Should one of them show up for some reason "in the trenches" with patients and physicians, he or she could only display merely "counterfeit courage," rather like that of a non-combatant trying to hob-nob with real soldiers—rather like present-day journalists 'embedded' in a military unit. The physician, on the other hand, "is never a mere observer" but is precisely on the firing line: accountable to, and held accountable for, patients. Philosophers may have their rightful place, but it is not in the clinic (much less the laboratory). Such academics and scholars belong, if anywhere, only on committees or panels deliberating policy—or, possibly, in classrooms educating students (though I wonder whether Siegler would have included medical students).

Although bioethics, as the broader field came to be called, has since become something of a veritable growth industry,[1] the issues raised in the early days continue, if anything, with even greater intensity even though so-called ethicists can now be found in most every major hospital in the United States and many other countries. Not even the fact that some physicians have begun to join the ranks of ethicists has done much to change this landscape; if they 'consult' in ethics, it is nonetheless the fact that they are physicians that legitimizes their presence. And, as emphasized earlier, to expect physicians (or patients/families) to be capable of conducting deliberative, probing conversations about the respective fundamental and often contentious sense of worth each participant may hold, not to mention hopes, fears, or issues of trust, seems at best unrealistic. An essential component of these moral conversations is concerned not only with the patient and family, but with the physician's own professional responsibilities and personal ethics—even, at times, with institutional values and social norms.

The key challenge is nevertheless perfectly clear. It does not concern, as was thought even in those earlier days, the curious notion that ethicists are, or take themselves to be, "moral experts" (Beauchamp 1982; Noble 1982, pp. 7–9, 15). Martin and Siegler had it right, it seems to me, for at the heart of that discussion was

---

[1] It is not insignificant to note that from meager beginnings in the early 1960s in the United States, this 'field' quickly became an international phenomenon of the first order: often required in some form or other by hospitals, physician groups, even governmental agencies, world-wide. A stunning development in the essentially socially conservative field of education alone, this development is all the more remarkable for its general acceptance in public arenas everywhere (Pellegrino and McElhiney 1981).

and remains another question, rather more difficult to confront: what, in the end, *is* the 'humanist' all about? Woven through that tapestry is a curious and even bizarre thread: can a humanist or philosopher really be a clinician? More forcefully, how can such persons possibly be held accountable? What, after all, do they *do*? Whether regarded as expert or not, in a word, when the philosopher or theologian do whatever it is they do, is that something for which they can be understood as somehow *responsible*? The discussions of the past two Chapters clearly stand in need of serious examination.

7.3   The Case Against 'Ethicists'

What I have in mind is quite apparent in a particularly abrasive example of the dispute (Ethics Consultants 1993, p. 2). It is one that I think put the concerns of many health professionals in a wonderfully concise way. In an article that provoked a quite lively discussion, but without any apparent awareness of the brief history I've traced in this book, the attorney, Giles Scofield, takes it simply for granted that many if not most of those who accept the idea and role of 'ethicist' invariably take themselves to be "professional experts" in ethics (Scofield 1993). While such persons of course have the appearance of being user-friendly, as it were, he believes that the truth is just barely concealed beneath that guise: a nefarious urge for power, authority, and legitimacy; that is, humanists or philosophers who call themselves 'ethicists' are merely in pursuit of the gold coin of our social domain, recognized "professional status."

To make his point, Scofield considers what he assumes are the four "basic elements of a profession" (Scofield 1993, p. 417). (1) Every true professional possesses "a specific body of esoteric knowledge." Ethics consultants do not possess anything like that, and even if there were such knowledge, (2) it could not be applied in the manner of any true professional, that is, "in an objective, reliable, and 'scientific' fashion" (Scofield 1993, p. 418). Indeed, his criticism gets into high gear at that point rather quickly. If we consider that the "epistemological foundations" of ethics consultation are surely dubious at best, that results of "ethical reasoning" are inherently variable and clearly open to bias and even bigotry, are unpredictable and non-reproducible—then it is obvious, to him at least, that the very idea that ethics consultants possess scientifically evident and professional knowledge is offensive, even outrageous.

(3) Professionals must be capable of self-regulation. But one can hardly expect this burgeoning field of "moral experts" to be professional in this sense. Contrary to that, in fact, there is as yet no trace of anything remotely like a professional code. Combine that with an historical record that makes plain their inability to conduct genuine internal self-reviews—the dismal record, in fact the complete absence, as he sees it, of any effort to police scoundrels and ineptitude in the field of ethics or ethics teaching—and it is clear that this field fails miserably to fulfill criteria of a genuine profession. Not only that, but, he argues, there is the well-known resistance

by those in the humanities to external scrutiny. Accordingly, he asserts, thorough skepticism (if not outright cynicism) is surely justified.

(4) Genuine professionals, furthermore, both individually and collectively act "in the public interest" generally, and for the benefit of patients or clients specifically. He then considers but swiftly rejects the notion that ethics consultants, despite everything, might be motivated to act generously, in the "public interest," out of altruism and not self-interest. For all the sworn hearts and raised hands, however, the very idea flies in the face of political realism and what Scofield believes is their real agenda: power and authority, not to mention salary and prestige—hardly indicative of serving the public welfare. So, he asserts (without evidence) that however noble is the spirit of traditional ethics, that is not necessarily transferred to its practitioners and, in the case of those who pursue ethics within medicine, this spirit of generosity is simply absent.

As if all that were not enough, he argues finally that (5) every genuine professional is and must be an expert in the field from which the profession grows and to which it is ultimately responsible. Therefore, he asserts, to be an ethics consultant is quite obviously to presume expertise in ethics. This means, for him that such persons are engaged in "applied normative ethics." The "ultimate problem" with that endeavor, however, is that each ethicist avows different and, one supposes, incommensurable claims,[2] none of which can "be true in a pluralistic, democratic society founded on the belief that each person is the moral equal of every other" (Scofield 1993, p. 423).[3] In such a society, there simply cannot be *any* moral experts, and anyone who presumes to wear that mantle is "essentially antidemocratic," his or her "claim to 'help' others is nothing other than a latent assertion of power and authority" (Scofield 1993, p. 423).

It goes without saying, of course, that Scofield would never want to find himself in some forlorn intensive care unit with little more than tubes and plugs to remind observers of his humanity, only then to suffer the bedside appearance of some "moral expert!" His rather colorful disdain to the side, what's he really troubled about? Whether cloaked in his undefended and poorly defined adherence to "democratic pluralism," or the more straightforward attack on the ethics consultant's presumed lust for "professional" status, the passion of his plea seems to me unmistakable. Beneath his taken for granted *trust* of "professionals" is his undefended presumption of a profound *distrust*: the very idea of a "moral expert" is repulsive. Which is to say, he simply asserts that it is incoherent that any philosopher or other humanities scholar (no mention is made of lawyers, though

---

[2] Although not cited by him, one hears faint echoes of Alasdair MacIntyre's analysis of the moral confusion of our times. (MacIntyre, *After Virtue*, 1981).

[3] Of course, even Scofield would have to make a list of exceptions to this typical sort of claim: no children, no one believed to be mentally disabled, etc., would be admitted into this charmed circle, not even by the most devout egalitarian. And, this lands him in the same paradoxes already familiar to most of us: how 'to draw the line' demarcating inclusion and exclusion? Scofield neither raises nor hints at how to handle the problem he himself presents.

there are quite a few who have entered the ranks of we 'rank' ethicists) can possibly be held accountable for words or actions in clinical or research settings.

Scofield takes it simply for granted that every one involved as an "ethicist" is a theorist engaged in "applied normative ethics." Although he never wonders what the place of any "theory" might be in medicine, he apparently believes, but again without giving reasons, that there simply is no place for "ethics theorists" in that world. For that matter, none of the ethicist's claims, given the characteristically severe disputes among their ilk, could possibly "be true in a pluralistic, democratic society founded on the belief that each person is the moral equal of ever other" (Scofield 1993, p. 422). So, each individual is his or her own moral expert, and any claim to the contrary is mere pretense, a sham.

He nonetheless tosses a small bouquet, for there is some "value to others" in what clinical ethicists know and have been doing before they got seduced out of their classrooms: they educate. Which, clearly, is quite an odd remark, as it would only keep in front of young and gullible students what he fervently wants evicted from the bedside and case-conference room—as if unadorned ineptitude and cupidity at the latter were not quite enough to undercut the former. Here, he merely reaffirms the old notion of the utter irrelevance of philosophy, since, like many people, he thinks that nobody is harmed in college courses—nor, for that matter, helped very much.

7.4   Aspects of the Dispute

Scofield was quite delighted with his assault—many of whose shots are doubtless quite cheap—for it in any event provoked several critical responses. He clearly thinks this fact gives his tirade some legitimacy. And, he thinks this meant that "I struck close to home," as he avers in his later reply to his critics (Scofield 1995).

In fact, however, most of his shots hinge on the very narrow and unexamined assumptions I just noted, many of which are clearly pointed out by John Fletcher (1993), Albert Jonsen (1993), Christian Lilje (1993) and Donnie Self (1993) in their various replies. Oddly, or perhaps not, he ignores the only respondent, Judith Wilson Ross (1993) who seems to agree with him, at least about the main role ethicists should play. She contends, "education is the function in which ethics consultants engage most often and persistently." But Ross goes Scofield one better, for placing clinical ethicist into what are essentially academic endeavors in the hospital, "is puzzling at best and impossible at worst" (Ross 1993, p. 445). Not only do ethicists come on that scene with "precious little endorsement from anyone in significant authority" (Ross 1993, p. 445). Beyond that, it is evident to Ross that such a hospital role carries with it reasonable expectations by health professionals that are at odds with the role of simple "educator." Hence, the fact that ethicists invariably appear as "offensive intruders trying to tell other people how to do their business" (Ross 1993, p. 446) is not in the least surprising.

Accordingly, Scofield's cynicism seems if anything underscored by Ross: "I think that clinical ethics consultants are up to something else," (Ross 1993, p. 445) she alleges. Nor does it seem at all difficult for her to figure out what they—more charitably, many of them—are up to under the guise of the role of "hospital educator:" the ill-concealed effort to create a new role and profession, "beeper ethicist," who responds day and night to "ethical crises." The whole thing, she believes, is little more than an oxymoron, whether conceived as hospital educator, beeper ethicist, or clinical ethicist. Ross thus clearly thinks that Scofield is very much on target, since, for the ethicist to be effective in the hospital environs surely means that they look and function like doctors, "as crisis managers. . .in acute and interventionist ethics," (Ross 1993, p. 447) a role better left to those more accustomed to such crises (one hears Siegler and others, though unmentioned, in the background of her argument).

If Ross is nonetheless more tempered in her assessment of clinical ethics consultants than Scofield, she nonetheless agrees with him in most particulars. But it didn't apparently dawn on her that Scofield's acid comments betray little knowledge or understanding of the actual nature and demands of clinical contexts or of ethics consultation, nor is there much in Ross' commentary that reveals any significant experience on her part in these settings. In fact, both of them, for at best obscure reasons, attack clinical ethics on essentially political grounds—and to some extent pedagogical ones. Neither, however, shows any sense of the substantial differences between politics or education at the bedside and what might be required, say, in ethics committees or policy deliberations. Neither betrays any relevant experience in what they nevertheless severely criticize. Neither, to be blunt, knows whereof he or she speaks.

It is of course perfectly true, as I've argued at length in previous Chapters and elsewhere, (Zaner 1988/2002) that federal or state regulations, say, are often among the issues that help configure a particular clinical encounter—the 'Baby Doe' and, in some states, 'assisted suicide' enactments are obvious examples for some clinical situations. At the same time, however, it ought to be just as evident that the political considerations at the macro level are quite different from those presented in the micro, bedside, context. There is no way, surely not the simple way presumed by Scofield, to go from the one to the other.

The same must be noted regarding education, for while it is obviously true that educating in a college class is quite different from educating at the bedside or in a clinical case conference—as I've already discussed earlier—both Ross and Scofield equivocate. In the one case, Scofield's bouquet to philosophers as educators, as noted, simply presumes without evidence that ethicists should depart from the bedside and return to the classroom. In the other, Ross apparently thinks that the move from the classroom to the "impossible" (or at best "puzzling") function as "hospital educator" frankly opens the irresistibly tempting door to "beeper ethicist," seducing academics to play like doctors and along the way flatly deceiving health professionals and patients/families.

But neither Scofield nor Ross seems in the least bothered by the pointed and wholly pertinent questions Dr. Martin put to the 'humanities' long ago: what, exactly, are they all about and, echoing his lament, if they are in any way helpful, how can their concerns be translated into clinical settings so as to help bring about 'the new physician' that so concerned him and others? Both Scofield and Ross would presumably agree with Lilje's ironic comment about keeping academics in the classroom: "in this least dangerous setting, ethics did not harm much, nor help much either!" (Lilje 1993, p. 440). But is that true, any more or less than it might be true about ethics in clinical settings?

Difficult and puzzling though it may be to one who finds classroom lecturing about ethics somewhat problematic, what reason is there to suppose that providing an opportunity for a patient to discuss her understanding, hopes, and fears candidly is unhelpful to that patient, or that the patient learns nothing from such a conversation? Furthermore, while it may be true that some ethicists might indulge in "a latent assertion of power and authority" at the bedside—though, even if Scofield is correct, it is surely open to question whether and how his own patently political interpretation bears on the dimensions of ethics—such assertions are in any case equally possible in the classroom.

In any event, one can only wonder how Scofield himself could ever know if or when there really is an "assertion of power and authority" by any so-called 'ethicist', latent or not, without actually going to the bedside himself and either carefully observing what goes on or, better yet, doing what he regards as illicit when practiced by others. More than that, he seems blissfully unaware that what he warns against at the bedside can and does go on in the classroom, not to mention that such "assertions of power and authority," just as likely occur on the part of doctors and nurses as by ethics consultants—if he's right about the hospital then he must also be right about the classroom. In either case, while his words trumpet warnings, they have little backing or warrant; he simply does not know, as Jonsen and Self also suggest, whereof he speaks.

Consider two strands in this turbulent stream, then. On the one hand, Scofield says that ethics consultation is misconceived because it doesn't fit into his understanding of the proper political scheme, that is, a pluralistic democracy where anybody's moral opinion is as good as any other. Which, however, simply assumes a ready translation from the political norms of society to the more fine-grained relationships at the bedside. On the other hand, neither Ross's nor Scofield's views provide an account for how to understand and assist individuals in clinical or in any other specific, situational encounter. Either there are no ethical problems at the bedside, then, or else the only way to consider them consists of letting physicians handle such problems—even though they can hardly claim any more insight into ethics than Scofield can claim knowledge of, say, physiology.

All things considered, it seems quite clear that Scofield's, and possibly Ross's, real target is not at all the ethicist at the bedside—however puzzling this is to both of them—but rather John Fletcher and the controversial fight he was at the time leading to certify, license, and regulate the burgeoning field of clinical ethics. What is needed to counter the abuses Scofield fears, Fletcher believes, is uniformity—of education, training, role definition, and formalized accountability by ethics committees (Fletcher 1993, p. 432). Although Fletcher wants to "put to rest unreal discussion of a 'profession' of clinical ethics," (Fletcher 1993, p. 433). Scofield is quick to point out that is precisely what Fletcher and the Board of Directors of the Society for Bioethics Consultation were in fact seeking. If anyone retains any doubts about that, Scofield gleefully asks that we merely consider that, as the President of the S.B.C. has stated, one of S.B.C.'s goals is to "maintain the S.B.C. role in the process of professionalization of bioethics consultation" (cited by Scofield 1995, p. 226).

Since that is the real point of his critique, Scofield feels at liberty to lampoon Fletcher's call for "regional bioethics networks"—a proposal that is, Scofield recognizes, a plainly political and economic move to corner the market—presumably for the S.B.C. and its accomplice, the Society for Health and Human Values, in the task force to create just such standards (Scofield 1995, p. 225). Beneath Scofield's evident suspicion is an obvious distrust over what he believes is, perhaps correctly, patent political maneuvering among the very 'professionals' who are yet leery of 'professionalism'.

Fletcher, too, worries about the field to which he has devoted many years, especially about the possibility of charlatanry and that the field risks being little more than snake oil, rather than genuine ethics. Fletcher's proposal, however, not only runs loggerheads with the *Realpolitik* that worries Scofield, but, without putting too fine a point on it, flies in the face of academic and theoretical reality—the deep differences, divisions, even divisiveness in those quarters are legendary. Not only are such discordances unlikely to be curbed or healed by appeals to some sort of "pluralistic democracy," but such an effort would come at too high a cost, for to mute or suppress that diversity would be to bottle up precisely what makes a variety of views most valuable. Moreover, the history of professions gives little reason to expect effective, willing self-regulation and certification standards. Indeed, the very idea of any sort of imposed uniformity, much less orthodoxy, among those in ethics seems if anything utter anathema and contrary to the free, candid flow of ideas so prized in and essential to their discussions.

Just that prospect, on the other hand, is what deeply concerns Jonsen, when he emphasizes how wrong-headed is Scofield's[4] understanding of clinical ethics: there simply "is no orthodoxy" among clinical ethicists but rather "considerable heterodoxy" (Jonsen 1993, p. 435). Of course, that may be but another reason for its critics to resist ethics consultation: if pluralism or heterodoxy is the proper name, doesn't that make the very idea of "beeper ethicist" all the more troubling? After all, if ethicists want to be 'like' doctors, isn't it only reasonable, even imperative, that when ethicists are asked to consult, doctors and patients should reasonably expect something like consistent appraisals and recommendations, no matter which ethicist arrives at the bedside? But if ethicists are, as Jonsen says, so different, doesn't that of itself shatter the very idea of ethics consultation?

---

[4] I presume, too, that this view implies that Jonsen is opposed to Fletcher's proposal.

7.5   So, What in the World Is 'Clinical Ethics Consultation'?

So where does all this lead? I, for one, am left with much the same hollow dismay and deep discomfort—and the same questions—I faced when I first came onto this scene in the late 1960s. If anything, I am even more deeply troubled, Jonsen's favorable reference to me, or Scofield's neglect, notwithstanding. For I must confess that I have been daily engaged for many years in precisely what Scofield lambastes as ill-begotten: a great many consultations about ethical matters with patients and their families, friends and surrogates, as well as physicians, nurses, social workers, chaplains, therapists, even (heaven forbid!) hospital managers and CEOs. Never mind that nothing of what I've done even remotely resembles anything of what Scofield or Ross assert.

And yet, I hesitate. I sense neither that I have all along *really* been "up to something else," nor that I've been lusting for power or authority, nor that I am antidemocratic, nor that I possess an esoteric body of knowledge that others should (must?) respect and call upon, nor even that I am somehow better or more 'expert' than anyone else. To the contrary, as I've made a point by emphasizing many times, I honestly believe that I've been a privileged witness, time and again, (Zaner 1994, 2004) to astonishing insights into what the moral order is all about, thanks especially to those people who invited and allowed me into their lives to listen and question, perhaps at times even to talk, as they struggled to make sense of what had happened to them, the decisions they had to face, ultimately, what their lives were all about.

To put Scofield's, Ross's, and others' concern as unflinchingly as I can: where do I—or Jonsen, Self, and the others engaged in some form of ethics consultations— get off anyway? What is ethics consultation all about, if not what critics like Scofield or Ross allege? What, bluntly, am I up to? Are we not 'experts,' if only in the amiable, unassuming sense Jonsen believes we ought to be? If not, then what are we? And why has so-called bio-ethics had such a show of popularity and growth over the past two or three decades?[5] Is Scofield privy to some sort of special insight closed to the rest of us, even though he's never done nor even observed the thing he's so anxious to criticize?

One could take the common tack of those commentators invited to respond to Scofield: embrace the concept of 'moral expert' but carefully lay out, as does Jonsen, the equivocation in that (Jonsen 1993, pp. 436–38). As Jonsen puts it, the notion of 'expert' refers to "the consultant's experience and skill issues in advice," (Jonsen 1993, p. 437) and that means, not at all the sort of craving for power and authority that worries Scofield, but the rather more gentle and "humble expertise"

---

[5] Especially acceptance in the world of clinical and academic medicine. I recall being stopped in the hall by the Dean of Medicine at Vanderbilt, not for a question for so he could observe, somewhat slyly, that 'ethics is the first new program at Vanderbilt in more than 36 years!' To his subsequent wink and smile I could only respond, 'What? I'm a 'program' and not just a professor?'

Jonsen believes is "a small blessing in the confusing and conflicted world in which we live" (Jonsen 1993, p. 437). True, Scofield scoffs at this very effort:

Ethics consultants are hardly mere advisors...[for] just as physicians can "get" patients to consent, ethics consultants can guide discussions so that others will agree to a recommended course of action as if they had thought of it themselves...thus, the ethics consultant's claims to "help" others is nothing other than a latent assertion of power and authority. (Scofield 1993, pp. 422–23)

To which Jonsen in turn scoffs:

New tyranny, indeed! It is equally possible to describe the advice of a consultant as liberation: persons whose minds and emotions are congealed by confusion, ignorance, or distress can be unfrozen by thoughtful and informed advice. Their ability to choose responsibly flows freely again. (Jonsen 1993, p. 438)

Jonsen has by far been the most explicit in describing what, at least, he does when he 'consults.' Setting aside his (as I think) quite justified rejoinders to Scofield's complaints, he takes leave to point out that he may be able to "assume that I am among the exemplary and responsible [ethicists], because Scofield does not cite any of the many articles I have published about ethics consultation (although I do not call myself a consultant)" (Jonsen 1993, p. 436).

7.6 Jonsen's View, My Criticism

Jonsen, like others of us, constantly lives with the "perennial" question, "is there any such thing as ethical expertise?" and responds to his own question, "I am an ethical expert," which is (clearing up Scofield's equivocation) a quite "humble" claim. Jonsen cites the OED to help clear up the issue even more: an "expert" is "one who has gained skill from experience," and "experience" is precisely "the state of having been occupied in any department of study or practice, of affairs generally or in the intercourse of life, the extent of which, or the length of time during which one has been so occupied, the aptitude, skill, judgment, etc. so acquired." The next step is at hand, since he (and a few others of us) have been so occupied and engaged "in the 'practice' of clinical consultation" for many years, and therefore we and he "have acquired...some aptitude, skill, and judgment" (Jonsen 1993, p. 436).

I might note here that Scofield completely ignored my own numerous writings on these matters, perhaps because none of them conforms to his claims about ethics consultation, pluralistic democracy, etc. He also ignored Jonsen's many writings; I will not. Rather than try and rehearse my own views of these matters—which I've already done at length here and elsewhere for the past four decades—I want instead to look carefully into what Jonsen has said. This critical examination will help me formulate my views even more clearly.

To understand his view, it is necessary to return, this time more directly, to the notion of casuistry he and Stephen Toulmin are well known for having resuscitated from the many abuses it has long suffered. "Casuistry," as they conceive it in their book, is

> the analysis of moral issues, using procedures of reasoning based on paradigms and analogies, leading to the formation of expert opinions about the existence and stringency of particular moral obligations, framed in terms of rules or maxims that are general but not universal or invariable, since they hold good with certainty only in the typical conditions of the agent and circumstances of action. (Jonsen and Toulmin 1988, p. 257)

The pertinence of this for clinical ethics consultations is cleared up a bit in his response to Scofield:

> The expertise of clinical ethicists arises from their encounter with many cases, among which they can see analogies, and from their educated understanding of the various meanings of rules and maxims (e.g., have respect for patients, do not kill innocent persons, do no harm, etc.) that are referenced in such cases. Expertise results from experience: seeing many cases over time and noting their similarity and dissimilarity. (Jonsen 1993, pp. 436–37)

Jonsen points out that the practice of casuistry was not the central concern of his and Toulmin's study; they "only laid the groundwork for so doing" (Jonsen 1991, p. 297). In another essay, however, Jonsen takes that step, proposing an interpretation of Aristotle's *phronesis* as the basic clue to the sort of casuistic reasoning at work in clinical ethics consultations. "The casuist," he argues there, "will be able to scan or parse the case, revealing its structure of claim, maxim, grounds, rebuttals," (Jonsen 1991, p. 306) or what he terms the "morphology," "taxonomy," and "kinetics" characteristic of casuistry:

> Casuistry will be able to locate the case in a taxonomy of cases, recognize the similarities and differences and appreciate the shift from moral certainty to moral doubt. Above all, casuistic reasoning is prudential reasoning: appreciation of the relationship between paradigm and analogy [i.e. taxonomy], between maxim and circumstances [i.e. morphology], between the greater and less of circumstances as they bear on the claim and the rebuttals [i.e. kinetics]. (Jonsen 1991, p. 306)

Marvelously succinct, this essay makes evident what he means by the aptitude, skill, and judgment that come from experience in dealing with analogous 'cases.' The casuistic method, he believes, gives the ethicist the exact tools needed for identifying ethical issues, clearing up confusion, and analyzing clinical situations so as to provide often much-needed 'advice' to those who must make choices and decisions. The casuist, thus, acts to "untangle the elements and rearrange them in a pattern where the participants can see their choices with clarity" (Jonsen 1993, p. 437). The example he gives in his response to Scofield is straight to the point:

> In the majority of cases in which I consult,[6] this sort of expertise is engaged. People want to learn what a "natural death act" is or how to interpret its language. They have not thought about the use of nutrition and hydration, cardio-pulmonary resuscitation, or persistent vegetative state until faced with dire decisions. They want to talk through the problem. The department of study and practice in which I work encourages me to think constantly

---

[6] Despite his disclaimer of the term, quite clearly Jonsen *does* regard himself as a "consultant".

about these things and to learn from others (including Mr. Scofield) who also think about them. (Jonsen 1993, p. 437)

The point should not be ignored: "some who have attempted to learn about a problem can share with others who have not yet encountered that problem or who may not have had the leisure or guidance to reflect upon it" (Jonsen 1993, p. 438). Thus, what drives Jonsen's "practice" *is* a kind of "expertise," one which he thinks should be shared with others who have not (as yet, only minimally, or not at all) had the requisite experience; altruism does indeed motivate him. Contrary to Scofield's nasty remark, Jonsen is hardly a "sneaky ethical fascist" (Jonsen 1993, p. 436).

Jonsen's sensitivity, humility and concern regarding those struggling with these difficult decisions are evident and cannot be gainsaid. Yet, I venture to suggest that Jonsen is nonetheless remarkably insensitive to a significant issue. What, as he sees it, does the ethicist, so-called, actually *do* when engaged in consultation? I do not mean what does Jonsen *say* he does in clinical situations when he is back at his desk; I mean in what does his "consultation" actually consist when he is at the bedside face-to-face with a patient and family members?

Jonsen seems to me candid and unequivocal: *he* "talks," at least for the most part, striving to help other people through the complex and difficult issues they face and who typically have had little experience, doing this "in an orderly way, pointing out salient features, revealing confusions, representing different interpretations. ...." Thus, when he is asked for help by a patient or family member, as he like others of us inevitably are, Jonsen says, "I often have an answer that will inform a decision." That answer *should* be "an expert opinion... [for it is one] shaped by reflective encounter with many similar cases and by dialogue with peers" (Jonsen 1993, p. 436).

To be sure, he advises that any "ethicist worth his or her salt knows, as a matter both of theory and of practice, that moral decisions are made only by the persons whose life, conscience, and responsibilities are engaged by the choice" (Jonsen 1993, p. 437). But *for what*, then, is Jonsen himself—like the rest of us—*accountable*? What is his, and our, *responsibility* in these situations? It would appear that, as is even clearer when he discusses the casuistic method, his aim is to share his expertise with them, but at no point is there a discussion, or even apparent recognition, of the ethicist's own experiences and responsibility for whatever it is that is "said" and "shared." Is it, then, that the ethicist only 'talks' and those who listen can, if you will, take it or leave it?

7.7 Responsibility in the Clinical Encounter

Whatever dialogue there is, so far as Jonsen actually discusses the matter, apparently takes place only "with peers," for there is not a word about any such dialogue, not even reports of actual conversations with those who ask for help. Nor is there a word about conversations with the other clinical professionals also involved:

nurses, social workers, and others. For that matter, Jonsen does not apparently learn much of anything from those with whom he "shares" his "talk;" *he* talks, *they* listen and hopefully learn. Is the only business then of ethicists to "talk" when asked, merely to "tell" others what the ethicist has learned from experiencing the "many questions, problems, cases, and opinions" in the "intercourse of life?" Does the ethicist not *listen* to those in dire straits, or learn nothing from them *while* they are in the midst of things and talking to him as they try their hardest to make sense and reach decisions? And, even if the ethicist "talks," must not that be talk for which he is *responsible* and *accountable*? To be as fair as possible to Scofield, among his deepest concerns seems to me to be precisely this question. It is an excellent question, but it is one to which Jonsen has no apparent response, and Scofield little more than sarcasm and innuendo.

I must again return to the question that, it seems to me, underlies these concerns, and has long been at the heart of my own understanding of ethics consultation. The main flaw in my earlier effort to make sense of these matters is that I did not (though I certainly meant to) give sufficient emphasis to what it means for those of us involved as ethicists to be truly *responsive to* the people who themselves confront the constraints and issues buried within clinical situations, and *responsible for* what we do and say. We have no convenient moral calculus to instruct us on what any set of moral principles requires of us in these concrete situations. Moreover, we might do well to acknowledge Hans Jonas' stern warning that we are

> constantly confronted with issues whose positive choice requires supreme wisdom—an impossible situation for man in general, because he does not possess that wisdom, and in particular for contemporary man, who denies the very existence of its object: viz., objective value and truth. We need wisdom most when we believe in it least. (Jonas 1974, p. 18)

Possessing neither convenient calculus nor supreme wisdom, what can be said? Earlier in my involvement, I noted that there are at least three requirements. (1) It is not possible to know just which moral issues are actually presented by any specific situation *in advance* of discovering the actual circumstances of their occurrence. This must be learned, and only attentive *observing and listening* can accomplish that—a superb example of what Kurt Wolff analyzed as *invenire*: to come upon, to discover (Wolff 1976; Zaner 2006a). Discussing the physician's duty to listen to his or her patients, Howard Brody notes, "the patient who feels *listened* to in the first encounter. . .is far more likely to show a positive response to treatment." It is thus clear that the "single most important predictor of relief from symptoms was the report of the patient that he had had a chance to discuss the problem fully. . .." (Brody 1994, p. 82).

Brody's point holds as well for any clinician, including the clinical ethics consultant—who must, in my view, do anything and everything it takes to understand and practice the point. More generally, as I stated the matter in the previous Chapter, "moral issues are presented solely within the contexts of their actual occurrence," which requires rigorous observing and listening to those whose situation it is.

(2) A good deal more must be understood about medical and health professional *practice* than is currently thought necessary—their history, how and why currently prominent themes became important, and connections with the biomedical sciences, at the very least. I have emphasized as much throughout this book.

Furthermore, as Mark Bliton and Stuart Finder emphasized, there can be little doubt that the changes in the provision of health care will result in additional ramifications, as yet unclear, for the ethical analysis of clinical interactions and other transactions (Bliton and Finder 1999). Therefore, because moral issues are presented solely in clinical situations, then surely there is a serious obligation to be attentive observers and listeners as well to those who are engaged in and challenged by those issues.

(3) The point that needs special emphasis, however, is that anyone who dares accept the challenges of clinical consultation must be *responsible* for his or her deliberations, conversations, and recommendations (if any)—in short, the conduct of ethics consultation must be *responsible*, the ethicist *held* responsible. In a sense, the involvement of ethics consultants who remain at a distance from the 'bedside'— including those who deliberate only as members of ethics committees—can only be artificial, perhaps even counterfeit. To be remote from the concrete clinical situation within which alone are the pertinent ethical issues at all presented, is to guarantee remoteness from those very moral issues and themes as well as to those whose situation it actually is. With real people facing real moral issues, *a responsive ethics must be a responsible ethics,* as I expressed it then. It thus strikes me as particularly egregious to propose, as does Fletcher, that responsibility (and, one presumes, legal accountability) for ethics consultations be given over to ethics committees, for to endorse remoteness from the clinical encounter in truth obfuscates the core clinical issue of responsibility and guarantees, I think, what questions need attention by those whose situation it is.

Frequently—though by no means all the time—sick and injured people and their loved ones experience their clinical encounter as overpowering, as if they had been set upon by circumstances beyond their control. As Robert Hardy learned while conducting numerous conversations with different people about their, or their loved one's, illnesses or injuries, people want most of all to know what's going on, what's wrong and what can be done about it[7]; but they also want to know that those who take care *of* them also care *for* them (Hardy 1978). Overcome, they may sense an inability to continue to do or be without stopping and urgently (hopefully, also thoughtfully) dwelling on what has befallen them. They thus often find themselves, if not descending into (momentary or more lasting) depression, then deeply engaged by serious questions: asking, examining, challenging, doubting, reconnoitering, exploring, sounding-out, rummaging about, prying, peering, unearthing—

---

[7] It should be noted that just this seems in some instances the pathos of the illness experience: wanting to know while precisely that is just what may not be knowable. People living through terminal illness often display just this profound enigma: why is that I must die? Why does cancer kill—not just kill, but kill me? Sick people do want to know, but it may not be at all clear just what it is that they want, and need, to know.

whatever it takes to find out, to resolve: that is, to get *free-from* their not knowing and thus to become *free-for* the kind of knowing and understanding which can lead to decision, choice, action.

Within the illness experience, to question is to beseech, appeal, and seek response. Questions invite and, when serious, hunger for responses, and in a double sense. In their questions, sick people and their families and loved ones seek what is *responsive* to their not knowing and their need to know (what's wrong, what will happen, what can be done about it—to find out what 'speaks to' one's plight relevantly and truly). On the other hand, sick people seek responses that are *responsible* (that is, something that can be counted on, reckoned with, held onto as, at the very least, addressing if not resolving the issues posed to them by the illness). Thus, in asking questions the questioner in effect is opened up to whatever responses might be forthcoming, and in this sense the questioner is and must remain *available to* or at the disposal of whatever speaks to his or her condition, hopes, and life. When the questions are *appeals to other people* for help—whether for cure, care, or comfort—they are appeals for others to share in the questions, both to understand and to be understanding.

7.8   On Narrative and Dialogue

As Hardy quickly learned, to ask a sick person, or a loved one, something about their illness or injury is immediately to find oneself listening to a story; sick people can indeed be eager storytellers avidly seeking listeners. And, like any story, theirs often meander, twist and turn, are frequently incomplete, sometimes wind around so cleverly that the point at issue may be lost (Zaner 1994, 2004). Then, further questions must be posed, gently and with great care, for to avoid a central concern is often evidence that one does not care. This may be deeply held religious beliefs that have now come under question; or questions arising from a person's feeling an uncertain and brittle future; or, it may be, beliefs and values so deeply held that they have never be truly focused much less deliberated. Suppose one grants that such questions and appeals for help and understanding are genuine. What about the invited responses? Simply because the appeal is sincere, does not in the least mean that the response called for will also be genuine.

There simply is no guarantee that the one invited to respond—the clinical ethics consultant—will either respond at all, will make that type of response which effectively shuts off further conversation, or will be responsive and responsible in the way the sick person seeks. Nor can the patient in effect *coerce* a physician, nurse, social worker, or ethics consultant not only to answer questions but force them to be both responsive to the patient's concerns and be responsible within whatever course the conversation may take (as well as for what is said and/or done). If ethics consultation in some sense initiates a narrative within the ensuing clinical conversation or dialogue, as I believe it does, there simply are no assurances in advance that the dialogue will actually be responsive to the patient's or loved ones' actual needs—or that the one who initiates and conducts the dialogue is or will be willing to be held responsible. There are no guarantees against charlatans, Scofield's and Fletcher's complaints to the contrary notwithstanding. Or, as was already emphasized in an earlier Chapter, there is no guarantee that trust will occur on either side, patient or ethicist. For that matter, not even a diploma on one's wall guarantees that trustworthiness has been earned, even if it may prompt one to believe in and trust.

Clearly, to be responsive to, and to be responsible for are not the same, and should not be confused. One can, as noted, respond in such a way as to shut off all chances for dialogue. Furthermore, to act responsibly, the clinical ethics consultant must also *explicitly and deliberately take on or assume* the responsibility for what subsequently ensues, with the full realization that this act may not itself be understood or appreciated. And, so far as what is at issue is the ongoing course of clinical conversations, being and being held responsible have their place, it seems to me, strictly within those actual, ongoing conversations. Finally, precisely in view of that, there is no way, in the end, for any of the dialogical participants to know, definitively, just what is in fact involved in being responsible in the 'here and now' of the ongoing conversation. I may well fervently wish to be (or to avoid being) responsible; but what that actually means within the dialogue will never be completely clear.

Again: even if such patients and their loved ones genuinely seek, at times desperately, to know and understand, there may be no guarantee at all that anyone, the physician or ethics consultant in particular, will listen—or even themselves understand fully what's going on. The problem, if you will, posed by clinical encounters is the same as that at the root of dialogue: its authenticity lays not so much with the questioner (the patient, family, loved ones) as with the *listener*. The questioning, as it were, knows itself to be genuine, even while the questioner may not truly understand or be able to withstand the brunt of what is at stake. The problem faced by the person who is seeking help is that he or she cannot be assured or guaranteed about the sincerity or legitimacy of the response, not even about whether anybody at all will actually listen and be responsive/responsible. Furthermore, to complicate matters even more, despite all this, the patient—who desperately seeks to know what's going on and whether the professional can really be trusted to do the right thing—seeks this understanding from the very persons who, in their relationship with the patient and family, have the entire advantage over them. The patient is, as the Hippocratic text says, "all in their hands." The structural forms of the existential asymmetry of power are always present.

The issue faced by the clinical ethics consultant, in somewhat different terms, is precisely there: that listening (not unlike reading) can only take place at the invitation of some questioner (or writer) and in that sense must be freely engaged with the full awareness of the risks inherent to any such invitation. Listening can no more be forced than can one can compel understanding; to attempt to do so would be to vitiate the very possibility of response, hence is to undercut the questioning which invited response.

Nothing can bring about responsiveness, then, except the free choice of the one appealed to and invited—whether physician, nurse, or clinical ethics consultant. But, if the latter chooses to respond—and there is no way to force that either, even if one holds an actual position in an institution—then the intrinsic demands of dialogue come into play. In being responsive to the appeal, listening and attempting to understand the patient's story, the consultant thereby not only claims to be responsive (to the one appealing) but also responsible (for what is said and done), and is thereby opened up for further questions. The patient's story openly invites the listener's response. The patient seeks whatever is responsive to his or her concerns; when and if the invitation is accepted, such responses are supposed to be relevant, pertinent and true to those concerns. But whether they are, and are understood or not, requires further dialogue, further chapters in the narrative being gradually built up between the participants.

To be sure, the clinical ethics consultant's responses may, and often do, assume the form of questioning and, if so, the consultant is presumed to have been attentive, his or her questions to be pertinent, to pertain to the matters at hand raised by the patient or loved ones. The patient's story, if you will, is open to questions to help the listener understand and thus respond. In either case, responses are also, like the questions which prompt and occasion them, essentially invitations to further questioning and responding—to shared conversation—until the point is reached (if it is) when the dialogical partners are freed-from not knowing and not understanding and thus freed-for truth, that truth which in some sense 'speaks' relevantly to their concerns and perhaps even resolves the process.

Obviously, at any moment, the entire process of clinical conversation can break down; it can be betrayed, people can be led astray, the whole process can be aborted, and for many reasons: blunt refusal to talk, failed insight, sloppy thinking and talking, dishonesty, impatience. There is, if you will, always a 'Scofield' character lurking in the fringes of every encounter. Precisely because it can only be freely undertaken, once initiated, clinical conversation around the patient's story (founded on honesty and courage) is unavoidably exposed to failures. Failures, in turn, reflect on those who make them: they then stand convicted by their own free choice, for it is that which alone can initiate the dialogue and sustain its course in shared discourse, and which is among the core features of moral responsibility. In a sense, though always lurking about, no genuine conversation in the least needs such a character. To speak only for myself, I am already quite aware of the dangers, and many more than Scofield ever knew.

To engage in such clinical dialogue, moreover, requires that we not be naive. After all, as Scofield would doubtless be quick to point out, such conversations are clearly open to all manner of tricks, guises and disguises, cunning and cleverness— there are many ways to exercise the power and manipulation he rightly deplores. There is doubtless no guarantee that clinical ethics consultation can avoid violence, no way *a priori* to forestall this troubling issue—although in our times it lies at the heart of far more than the clinical encounter, as it infects most areas of interaction among people. You can no more guarantee the other's honesty than you can always certain of your own innocence.

But the ever-present possibility of violence—of violating a patient and loved ones, whether in word or deed—has another side. This, it seems to me, is an essential component of the moral order[8] within the clinical encounter: that, in a word, although one *can*, one *ought never take advantage* of those who are disadvantaged. Whoever may be involved in it, the clinical encounter is essentially *asymmetrical*, as I have emphasized many times. Not only does illness compromise the patient and loved ones, but, by seeking help from others who profess the ability (knowledge, skills, resources) to help, so does the relationship itself place the patient and loved ones at a distinct disadvantage. In this, as Arthur Kleinman noted, illness is *demoralizing*: it compromises distinctive human capacities (awareness, understanding, choosing, etc.); receiving help may be *remoralizing*, an enabling and affirmation of oneself (Kleinman 1988). Brody's point about the physician-patient relationship is apropos here: just as, on the physician's side, "there is a deeply rooted 'need to know'. . ." so there is on the side of the patient "an equally deep 'need to be known'. . ." (Brody 1994, p. 81) or in the terms I've used before, there is a need for the clinician, physician or ethicist, not only to understand but to *be understanding*, which is the heart of compassion.

Precisely because the appeal within the illness experience is of the moral order, the response it invites and seeks must be as well: what is requested is a morally responsive and responsible response. To take advantage of the multiply disadvantaged and vulnerable persons seeking and inviting help, accordingly, is morally to violate them. Although he probably did not intend it this way, Scofield's invective against clinical ethics consultation thus seems to me, to the contrary, to express the challenge and risk that clinical ethics consultants must squarely face at every moment.

Accordingly, the variety of tricks, devices, disguises, guises, and potential violations of relationships among people are of the very essence of clinical dialogue, they are at its commanding center. It was for just this reason that, in an earlier Chapter, I insisted that the core moral feature of the clinical event is the challenge made evident by the Gyges tale. At every moment of interaction with patients and their loved ones, as well as with health professionals, the clinical ethics consultant is, or ought to be, the constant reminder of the moral freedom which is required for such conversations to be authentic.

In a precisely parallel way, the challenge to be morally responsive is a challenge ineluctably haunted by the temptation instead to assume power, to manipulate, to take advantage. To encounter persons rendered vulnerable by illness is thus to encounter a moral challenge to respond, *with* its attendant risks of violence, the temptation to take advantage of the disadvantaged vulnerable person. In just this sense, clinical ethics consultation—listening and helping others to ask and confront themselves and the issues occasioned by their concrete circumstances—is and can only be an act which, freely engaged, is at the same time a potentially *freeing* act. It is, accordingly, an act whose prime characteristic is responsibility. In the face of

---

[8] This might possibly even qualify as one of those elusive phenomena, a 'principle'.

existential vulnerability, it is a staunch refusal to take advantage and thus is the free acceptance of being responsive and responsible and, in the end, is the steadfast effort to elicit and nurture responsibility by those whose situation it is.

7.9   Are There 'Experts'?

What is it to be responsible? That, it seems to me now—as it did even when, in 1981, I began my long adventure in clinical ethics consultation, trying to respond and be responsive to sick and compromised people within their specific circumstances—is the issue that is posed by the apparently interminable concern over 'expertise.' Jonsen is of course quite correct with his emphasis on experience. One does—I should prefer to say that one *may* learn from one's own experience, since there are people who seem incapable of that—but, it is also obvious, this is possible solely if one is disposed, is constantly and deliberately ready and open, to learning.

Learning surely requires a constant, careful cultivation of respect for the unique, the always *different* that each situation invariably presents and makes it what it is. Jonsen's primary emphasis on "paradigms," "analogies," and "similarities," thus, is very troublesome. That emphasis risks obscuring what must never be obscured: what marks out each clinical encounter, for all the tempting analogies with other such encounters, is precisely what *differentiates* it from every other encounter; it is the uniqueness of the narratives within each clinical event that is faced and only later, in calm reflection, are questions of similarities and analogies apparent or relevant. In clinical ethics consultation, *differences here make all the difference.*

Accordingly, if experience (listening and learning) within these encounters teaches anything, it is that one must never make the very moves that Jonsen seems most inclined, even anxious, to make: to focus first on the analogies and paradigms—which are exactly *not* what confronts one here and now; second, to conceive the ethicist's role as primarily sharing his or her 'expertise.' The latter may well be among the best ways to *miss* the very things that are actually going on. If one talks instead of listens, one fails to show respect for the very persons whose situation it is; it is, after all, their circumstances, their problems, their values, and their lives that are at stake, not the consultant's (or at least not in the same way). It is to them the ethicist must turn for insight; those whose circumstances pose the problems are, after all, also the ones from whom resolutions must come.

More than that, emphasizing talk instead of listening runs the serious risk of a double failure. By failing, or likely failing, to understand those persons and their circumstances in their own terms, in their own words and circumstances, frameworks or narratives, one risks as well the grievous moral failure of possibly aiding and abetting their misconstruing what they face and how they live. It should be all too obvious that one cannot observe and listen, no matter how willing and able, without actually being on the clinical scene and understood as legitimately invited into and involved in that scene.

But to be involved as an ethicist has its peculiarity—the frequently heard notion of 'intruder' or 'interloper' may then come to mind, and for understandable reasons. Put most straightforwardly, the moral listener-consultant ought never intrude but rather can become involved solely on condition that he or she is *invited* into the situation and conversation—or risk being precisely a mere eavesdropper and busybody. For this very reason, no one-sided 'case-presentation' (whether from physician, nurse, or ethics committee member) can possibly substitute for actual earnest listening to and substantial conversation with each and every participant, individually and, where possible, together. Furthermore, the invitation can only come from those whose lives or professions are most immediately at stake: the clinical participants who must make the choices and decisions, and live with the aftermaths.

While I recognize the deeper complications implicit to this theme of invitation, I want only to note that its basis is and can only be a form of *trust*. And this act is complicated, as already indicated. For now, it is important to recognize that there is a form of trust that underlies and textures the act of inviting or allowing the ethicist (or any other would-be listener) to be there and to listen, but it is equally important to recognize the continuing trust in him or her after the invitation is given and accepted, and conversations begin and continue.

To accept that invitation, furthermore, signifies the acceptance of *responsibility*: trust invites, even requires, responsibility in the complex sense delineated already. To accept such an invitation, thus, is to encourage and enable the participants to talk and discuss, to tell their stories candidly in whatever ways they wish, and that the ethicist will observe, listen, question, listen, encourage further details, and listen still again—in the concerted, focused effort to hear and, perhaps, even help give them their needed moral voice and courage to hear themselves in their own telling, as they are encouraged to probe ever more deeply into their own lives and circumstances and, ultimately, to take responsibility for what must be done and lived with.

7.10 Review and Over-View of What's Next

To take on that complex responsibility, finally, signifies at least three things. First, it is to make the hard, complex effort *to be responsive to* those who have allowed the ethicist into their lives. Second, it is *to be responsible for* what is then heard, done, and said. Having asked for and then (if offered) accepted the invitation, in short, one cannot then back out or try to cancel that act—not, surely, without powerful reasons that are openly shared with those issuing the invitation. Third, if patients, families, doctors, nurses, and others must somehow be or become courageous in their circumstances, so must the ethicist—which is the further meaning of a responsible and responsive ethics consultation.

Three points, then. (1) However tempting the lingo of the times—health as a 'product' to be bought and sold, physicians (and others) as 'managers' and gate keepers, and the rest—even slight acquaintance with the moral issues posed by clinical encounters should suffice to counteract the temptation to construe the problems facing ethics as merely minor irritations. For not only is the treasured Hippocratic moral tradition absolutely imperiled in these changing times, but the commercializing of clinical (indeed, human) life can only barter away what is left of our moral compass. There can be no price set on these lives, nor on clinical encounters and their essential ingredients: dialogue, talking and listening, questioning and responding, about what matters most to those who face and must eventually have the courage to make choices and decisions. None of that can be slotted into the problematic envelope of 'management.'

(2) The disputes over the idea of 'moral experts' have been amusing if also aggravating. In the end, they have served more to stir up dust than to shed much light on the hard questions of moral responsibility. As I have urged this for clinical ethics consultations, so I staunchly advocate it for the classroom as well—a point that is utterly absent from Scofield's argument and most of the responses to it, especially Ross's. The questions and themes at the heart of the acts of teaching and learning are much the same as that of consultation; failure at the one bespeaks little if any hope of its being present at the other. So, it is perfectly obvious that, as so many have remarked, the ethicist is an educator—but to leave it at that, unanalyzed and at best poorly understood, does not bode well for either act, educating or consulting.

(3) Perhaps the most neglected facet of clinical ethics consultation has been the curious failure to appreciate that the ethicist, however understood, does in fact get involved, and that involvement has its inevitable consequences on the ethicist him/her self. I am myself at stake in my clinical involvements; like it or not what happens in these conversations affects me. If others are helped or harmed, so is the ethicist, for the act of involvement is necessarily reflexive, it reverberates back onto the ethicist in distinctive ways quite as much as it has its own kind of affect on the other clinical participants.

Only something of what this signifies can be suggested at this point, although I it surely needs much careful study (Zaner 1981, 2012). That involvement, in a word, harbors a critical question for anyone in ethics: what exactly, if anything, is the 'ethics' of 'ethics consultation?' What justifies this act, my own decision to become 'involved' in all the ways consultants invariably display? Beyond that, focusing on the effect of consulting on the clinical ethics consultant, what exactly happens and ought to happen *to the consultant* in the act of consulting? How, echoing Sam Martin's plaintive plea, is it truly possible to 'humanize,' not so much the physician as, instead, the 'humanist'—"the one," he says, "who must help us all?"

I am fully cognizant, finally, that my insistence on narrative and dialogue has not been sufficiently elaborated or defended. I have gone some way in other writings, (Zaner 1999, pp. 99–116, 2006b, 2012) and I have also suggested something of the connections between dialogue and freedom—namely, that in those clinical conversations that are like dialogical engagements, participants *collaborate* in each other's freedom. The one who responds enables the one who needs to know freely to continuing seeking through questions; each recognizing, welcoming, the same by the other.

Suffice it to conclude here with a brief word about the fundamental moral challenge, and thus risk, of clinical narrative, conversation, and dialogue. It is this: that those who agree to engage in it face an ultimate test, trial, perhaps anguish, namely the actual prospect of having to live with the knowledge that one of them may well choose (thoughtfully or not) to pretend, even to let the conversation be trivialized or go unheeded, that, in short, dialogue may not occur. How one can come to accept that, and resist the then very real temptation to use force and coercion on the other, however subtly—that is perhaps the awesome challenge to one's own integrity and courage. It may be, in the end, that one's free act of acceptance of the responsibility, risks and challenges of dialogue are what alone enables the other to be free and responsible—and perhaps at the service of still others who also seek response.

References

Beauchamp, Tom. 1982. What philosophers can offer. *Hastings Center Report* 12: 13–14.

Bliton, Mark J., and Stuart G. Finder. 1999. The eclipse of the individual in policy (Where is the place for justice?). *Cambridge Quarterly of Healthcare Ethics*. In *Performance, talk, reflection: What is going on in clinical ethics consultation*, ed. Richard M. Zaner. Dordrecht: Kluwer Academic Publishers.

Brody, Howard. 1994. My story is broken; can you help me fix it? Medical ethics and the joint construction of narrative. *Literature and Medicine* 13: 79–92.

Callahan, Daniel. 1975. The ethics backlash. *Hastings Center Report* 11: 1–14.

Charon, Rita. 2006. *Narrative medicine: Honoring the stories of illness*. New York: Oxford University Press.

Ethics Consultants and Ethics Consultations, Special Section. 1993. *Cambridge Quarterly of Healthcare Ethics* 2(4): 417–448.

Fleischman, A. 1981. Teaching medical ethics in a pediatric training program. *Pediatric Annals* 10: 51–53.

Fletcher, John C. 1993. Commentary: Constructiveness where it counts. *Cambridge Quarterly of Healthcare Ethics* 2: 426–434.

Hardy, Robert C. 1978. *Sick: How people feel about being sick and what they think of those who care for them*. Chicago: Teach 'Em, Inc.

Ingelfinger, F.J. 1975. Editorial: Ethics and high blood pressure. *New England Journal of Medicine* 202: 43–44.

Jonas, Hans. 1974. Technology and responsibility: Reflections on the new tasks of ethics. In *Philosophical essays: From ancient creed to technological man*, ed. Jonas Hans. Chicago: University of Chicago Press.

Jonsen, Albert R. 1991. Casuistry as methodology in clinical ethics. *Theoretical Medicine* 12: 295–307.

Jonsen, Albert R. 1993. Commentary: Scofield as Socrates. *Cambridge Quarterly of Healthcare Ethics* 2: 434–438.

Jonsen, Albert R., and Stephen E. Toulmin. 1988. *The abuse of casuistry: A history of moral reasoning*. Berkeley: University of California Press.

Kleinman, Arthur. 1988. *The illness narratives: Suffering, healing and the human condition*. New York: Basic Books.

Lilje, C. 1993. Commentary: Ethics consultation: A dangerous, antidemocratic charlatanry? *Cambridge Quarterly of Healthcare Ethics* 2: 438–442.

MacIntyre, Alasdair. 1981. *After virtue*. Notre Dame: University of Notre Dame Press.

Martin, Samuel P. 1972. The new healer. In *Proceedings of the 1st session, Institute on human values in medicine*, ed. Edmund D. Pellegrino, 5–27. New York: State University of New York.

Noble, C.N. 1982. Ethics and experts. *Hastings Center Report* 12: 7–9.

Pellegrino, Edmund D. 1986. Rationing health care: The ethics of medical gatekeeping. *Journal of Contemporary Health Law and Policy* 2: 23–45.

Pellegrino, Edmund D., and T.K. McElhiney. 1981. *Teaching ethics, the humanities, and human values in medical schools: A ten-year overview*. Washington, DC: Institute on Human Values in Medicine, Society for Health and Human Values.

Ross, J.W. 1993. Why clinical ethics consultants might not want to be educators. *Cambridge Quarterly of Healthcare Ethics* 2: 445–448.

Ruddick, William. 1981. Can doctors and philosophers work together? *Hastings Center Report* 11: 12–17.

Scofield, G.R. 1993. Ethics consultation: The least dangerous profession? *Cambridge Quarterly of Healthcare Ethics* 2: 417–426.

Scofield, G.S. 1995. Ethics consultation: The most dangerous profession: A reply to critics. *Cambridge Quarterly of Healthcare Ethics* 4: 225–228.

Self, Donnie J. 1993. Commentary: Is ethics consultation dangerous? *Cambridge Quarterly of Healthcare Ethics* 2: 442–445.

Siegler, Mark. 1979. Clinical ethics and clinical medicine. *Archives of Internal Medicine* 139: 914–915.

Wolff, Kurt. 1976. *Surrender and catch: Experience and inquiry today*. Boston/Dordrecht: D. Reidel Publishing Co.

Zaner, R.M. 1981. *The context of self: A phenomenological inquiry using medicine as a guide*. Athens: Ohio University Press.

Zaner, R.M. 1988/2002. *Ethics and the clinical encounter*. Lima: Academic Renewal Press (Reprinted from: (1988)). Englewood Cliffs: Prentice-Hall, Inc.

Zaner, R.M. 1990. Medicine and dialogue. In Edmund Pellegrino's philosophy of medicine: An overview and an assessment, ed. H.T. Engelhardt. Special issue of *The Journal of Medicine and Philosophy* 15: 303–325.

Zaner, R.M. 1994. *Troubled voices*. Cleveland: Pilgrim Press.

Zaner, R.M. 1999. Performance, talk, reflection: What is going on in clinical ethics consultation. In R.M. Zaner (guest ed.), Special issue for *Human Studies* 22(1): 1–3, 99–116.

Zaner, R.M. 2004. *Conversations on the edge: Narratives in ethics and illness*. Washington, DC: Georgetown University Press.

Zaner, R.M. 2006a. It's the body that matters. In *The sociology of radical commitment: Kurt H Wolff's existential turn*, ed. Backhaus Gary and Psathas George, 155–170. Boston: Roman & Littlefield, Pubs, Inc.

Zaner, R.M. 2006b. A meditation on vulnerability and power. In *Health and human flourishing: Religion, medicine, and moral anthropology*, ed. Carol R. Taylor and Roberto Dell'oro, 141–158. Washington, DC: Georgetown University Press.

Zaner, R.M. 2012. *At play in the field of possibles: An essay on free-phantasy method and the foundation of self*. Bucharest: Zeta Books.

Chapter 8
Clinical Listening, Narrative Writing

8.1 The Several Ways to Consider Any Clinical Encounter

Over the past 25 years, I was privileged to consult on a large number of individual clinical encounters, during which I was seen, and served, as an "ethicist". At the same time, these and other facets of health care attracted my philosophical interests. Accordingly, a distinction became necessary so to enable keeping very different sorts of questions and problems distinct.

(1) To consult as an ethicist on a clinical event is to be invited by one or more of the main participants to assist in the situation by focusing on the moral aspects (problems, puzzles, dilemmas, etc.) of the individual circumstances and constituents (people, settings, conditions, issues, etc.) themselves, *for their own sakes*. In this respect, the ethicist's concerns are, like any consultant, strictly therapeutic: attempting, for instance, to help a couple understand what they face when they are confronted with a highly problematic pregnancy; to help, where necessary and possible, the providers understand the concerns and situation of patients and families; or to assist a family identify and consider the issues they must confront when continued treatment for one of their members is thought to be futile. Even when the ethicist writes about the situation, in chart notes or other ways to record its specific aspects, the focus is strictly on and for the sake of those with whom the ethicist has been invited to consult.

Whichever encounter it may be, what are the questions the participants themselves have to face, whether they want to confront these questions or not? How can they be assisted to face those issues clearly and squarely? What does the situation itself make it necessary for them to face and find some way or other to resolve, so that they may go on with their lives? More simply: what are the issues any specific clinical event poses, which decisions must be faced and what choices made, and with which aftermaths of decisions are they most likely to be able to live with? The ethics consultant seeks to help such people become more aware of and clear about

© Springer International Publishing Switzerland 2015
R.M. Zaner, *A Critical Examination of Ethics in Health Care and Biomedical Research*, International Library of Ethics, Law, and the New Medicine 60,
DOI 10.1007/978-3-319-18332-9_8

their own moral views so that they can more likely reach decisions commensurate with those views.

(2) On the other hand, such clinical encounters have many times moved me to reflect on them—whether to gain better understanding of moral life, of moral agency, or some other matter. Clearly any example whatever of clinical encounters, or any of their aspects or characteristics, can be reflectively considered in many respects strictly *as examples*—examples of clinical consultations, or some aspect of them, methodically considered in order to determine which themes are common to the range of examples, and what that may reveal regarding the questions which prompted reflection in the first place. In turn, these common themes may themselves be systematically considered in further reflective work, leading ultimately to a more embracing, more generalized philosophical understanding.

In this case, the philosopher's attention and focus is strictly on a particular encounter, *not* for its own sake, but rather for the sake of what is exemplified by and through it and others examples of such encounters. Such reflective consideration is most helpful to the work as a consulting ethicist, but must not be confused with it. Either manner of attending to a particular event, thing, person, etc., either for its own sake or as an example is possible for any phenomenon. How they relate and mutually clarify each other, as well as the specific nature of each mode of attention are issues, among others, that cannot be taken up here, but must be left for another descriptive analysis (Zaner 2012). What follows below is but one effort to become clearer about prominent features of the clinical event, and is based on my own experiences as a consultant on ethical issues in many clinical situations.

8.2 Encounters Are Context-Specific

Whatever the clinically presented problems may be, they are strictly problems facing the people whose situation it is most immediately—for instance, the expectant couple mentioned above and their physicians.[1] By the same token, the problems, alternatives, decisions, and outcomes, are strictly theirs as well. Any encounter presents its own set of issues, moral and other, and these are *context-specific* in the sense that working with and listening to such persons, helping them appreciate and advising them regarding specific issues needing resolution, and the like, requires a strict focus on the *situational definitions* of each involved person (Schutz and Luckmann I 1973). To understand a clinical encounter, there is nothing for it but to try one's best to get at the concrete ways in which the participants themselves experience and understand themselves and their circumstances, and endow its various components (objects, people, things, time, relationships, etc.) with meaning, in light of which they proceed to make decisions and act.

---

[1] As I note later, others—persons, professions, and institutions (with their departments, units, etc.)—are also invariably involved, albeit in different ways.

Thus in the situation presented by the problematic pregnancy mentioned above, (Zaner 2005a, b, 2007–2008) although the attending physician had told me that "abortion" was the "problem" needing attention, this was not in fact an issue— neither for the couple nor for the attending physician, since without mentioning this to each other they were each prepared to accept the possibility of early induction and fetal demise. When the dismal prognosis was mentioned and the couple seemed to become "angry," however, the attending broke off the conversation, as he thought they were angry with him for using the word "abortion."

With conversation at a standstill, one issue was obvious: to enable them to talk and listen to one another and, thus, to straighten out the different understandings in order to identify precisely what was at issue for each of those involved, so as to work toward a common understanding of problems, needed decisions, and, hope- fully, acceptable solutions. As was emphasized above in Chap. 6, in every clinical encounter, moral issues specific to the participants and their circumstances are presented for deliberation, decision, and resolution solely within the contexts of their actual occurrence. To find out and understand what's going on in any clinical encounter—what's troubling the people, what's on their minds, and thus to know what has to be addressed and how—requires cautious, attentive listening and probing of their ongoing discourse, conducts, the setting, and other matters presented as constituting this specific context. For instance, the couple's puzzle- ment about the meaning of "statistically significant"—a term used by the physician while discussing the results of several tests of the fetus—was central to what was interpreted (prematurely, as it turned out) as their "agitation" and "anger," and this indicated (at least in part) one important theme to be addressed. But these matters could not be considered in abstraction from the actual circumstances: what each person understood, what this led them to think about, etc.

I was invited into an already ongoing clinical encounter between the couple and their physician (and others: nurses, medical and radiological consultants, etc.). As I've emphasized earlier, physicians are in the nature of the case involved in a complex moral relationship with persons who, due to impairment (in a broad sense, including a difficult pregnancy) and to the relationship itself, are uniquely vulner- able, exposed to the power of those who wield the 'art' (*tēchne*). The latter are in turn under the obligation always to act justly and with restraint.[2]

In this respect, every physician (and other health provider) is as such focused exclusively on helping each patient under his or her care: diagnosing, outlining available therapies, and working with each patient to reach decisions most accept- able and reasonable to both physician and patient (and, at times, the family and/or loved ones). The patient (and family) is on the other hand focused on the his or her own condition and on doing whatever is necessary to be cured, feel better, or at least

---

[2] The Hippocratic virtues—justice (*dīke*) and self-restraint (*sophrōsyne*)—commonly accepted as key to the Western medical traditions come into play at just this point.

be helped as needed (perhaps, with palliative care, with pain management, and the like).

I have also emphasized earlier that neither patient nor physician (nor other health care provider) is focused on the relationship itself as the primary theme. To be sure, they are, as noted, obviously aware of that relationship; but reflecting on the relationship as such (for instance, with determining and assessing its asymmetrical structure) is not their primary concern. Patients are not philosophers, though philosophers, of course, are patients from time to time. Just that mutual relationship, however, is precisely the focus of the clinical ethics consultant and will, moreover, be a central theme for the reflecting philosopher when he or she considers examples of clinical events. This needs to be clearer.

8.3 Illness and Meaning

Experienced by the impaired person, the impairment is also interpreted by, and thus has meaning for that person. Others also experience and interpret the person's condition: family members, those in the person's circle of intimates (especially close friends and associates), persons in the wider social ambiance, but also the physician, nurse, and other providers helping to take care of the person. Hence, to speak of "the experience and meaning of illness," as many including myself have done, (Cassell 1973; Kleinman 1988; Pellegrino and Thomasma 1981) is necessarily to face a highly complex phenomenon—but this has not often enough been recognized.

Nor is this all. As Schutz has shown, every situational participant experiences and interprets the encounter within his or her own biographical situation: (Schutz and Luckmann I 1973, II 1989, pp. 243–47) typifications, life-plans, senses of self and others, undergirding moral and/or religious frameworks, etc. These encounters are socially framed by prevailing social values, as well as by written and unwritten professional codes, governmental regulations, hospital policies, unit or departmental protocols, etc.—any or all of which may and often do contribute to 'what's going on' in any specific case.

Cautious probing reveals that experience and meaning are still more complicated. Again, consider only a patient and her physician. She, like every patient, is *this* person, a *self,* and thus is essentially a reflexive being (Zaner 1981, pp. 144–64). Briefly, this signifies that the patient experiences and interprets her own problems or impairment. She also experiences and interprets the physician's conduct, physiognomic expressions, experiences and interpretations, including his experiences and interpretations of her (how she is thought to experience and interpret the doctor, her illness, etc.). And both she and her physician are, in the nature of the case, aware of, though not always focally attentive to this very complexity. In a word, the relationship is complex and reflexive: minimally, each experiences and interprets the other, their various expressions, their respective interpretations, and at the same time, within their relationship each experiences and interprets the relationship itself (Kierkegaard 1954; Zaner 2005a).

In the terms used by Aron Gurwitsch,[3] every constituent of a contexture is related to every other constituent (each is placed in some way with respect to every other), and vice versa; furthermore, each is related to the entire set of relations, as is the set itself related to each constituent. The relationships that constitute a contexture are, thus, at once reflexive (by referring to another constituent, the first relates to itself as what the other is related to; and vice versa) and therefore complex (a 'whole' is precisely that entire set of mutual, complex and reflexive relations).

For example, the pregnant woman in the encounter already mentioned not only experiences her pregnancy and her developing fetus,[4] but this experience is complexly textured by the ways in which she experiences and interprets what her physician (husband, and others) tell her.[5] Similarly, her physician experiences and interprets her words, expressions and gestures—for instance, he interpreted her "anger" as directed at him and his use of the terms, "statistically significant," and "abortion." In some respects, moreover, both of them experience and interpret the relationship itself. Regarding diagnostic data, for instance, she told me, "I know they're only trying to do their best" (i.e. she interprets the relation as "they are only trying to help"); and her physician said, "She seems to think we're being deliberately unclear" (i.e. the relation is "not going well"). But, as emphasized, the relationship itself—its characteristics and features—is not reflectively attended to by either of them.

To work as an ethics consultant is thus to be a kind of detective or, better, a type of literary interpreter: deliberately probing into the multiple situational 'texts' or ways in which the situational participants interrelate, variously experience, talk, listen, and interpret one another. The involvement of the ethicist is thus a *work of circumstantial interpretation* (both understanding, and being-understanding); reflection on this and other cases is a matter of *phenomenological explication*.

What has been pointed out, to repeat, emerges from considering the range of clinical encounters as examples; that is, from philosophical reflection. By contrast, clinical consultation (as opposed to describing or talking about it) is a specifically different kind of activity: its focus is on the effort to listen to and help the unique and individual in specific circumstances and for the sake of the individuals themselves.

---

[3] See, for instance, his *Field of Consciousness*, Duquesne University Press, 1964.

[4] By now, moreover, there seems little doubt but that the fetus also relates, in its own unique manner, to the womb bearing the fetus.

[5] As anyone will recognize, after even brief reflection, even a physician's raised eyebrows at a specific moment during a conversation with a patient, can be taken by the patient as highly significant.

8.4   What Is to Be Discerned

The stories patients and others tell invariably arise from and express themes intrinsic to clinical encounters—more accurately, from those encounters in which patients understood that much was at stake, much to be won (by 'successful' treatment) and much to be lost (when everything has been done and rescue, cure, or restoration is no longer possible). At the core of these clinical events is an encounter with our own, specifically my own mortality and the circumstances which make that especially exposed and exigent: questions of dying and death, loss and grief, and how people deal with them.

It seems true that most of us nowadays are only too well acquainted with the excruciating experience of having to deal with the death of a parent, child, spouse, close friend or other loved one. Those of us in our middle years have learned that we have no choice but to count on facing such situations: not merely to think and make choices about what to do for terminally ill relatives—parents, grandparents, children, or others—but to talk about them with strangers—nurses, administrators, physicians and, at times, social workers, chaplains, and still others—many of whom possess real power to control what will occur during those arduous, sad, and painfully extended moments of serious illness or injury, especially at the end of life.

It is not easy to for any of us to discuss such situations in any event; it is all the more awkward and difficult to talk about when someone else who is an intimate is faced with the extreme situation: someone close and dear is dying or faces severe compromise, is in great pain with relief only barely in sight, if at all. It is for most of us next to impossible when the individual in question is yourself, is myself. When faced with such situations, Ronald Blythe's terse comment about Tolstoy's *The Death of Ivan Ilych* comes to mind: that the character of Ivan reveals the "plight of a man who has a coldly adequate language for dealing with another's death but who remains incoherent when it comes to his own" (Blythe 1991, p. 10). Faced with the prospect of my own dying—say, on receiving a diagnosis of serious cancer—I may be struck dumb, without words or wits to withstand the onslaught of the unspeakable. But, it has also seemed to me, faced with the pending death of a loved one—wife, husband, child, sibling, mother, father—so are we often struck dumb as well. Facing our own not being, just as when we face the no-longer-being of a loved one, we know the profound inability of language to say it well, and for us to bear needed witness.

We rarely if ever have the right words at hand, if we ever do have them, to talk candidly about dying, loss, grief, profound sadness, fear, dread—the inner tremblings of the soul. Instead, we fumble and mumble, waiver and stall for time and still more time waiting for some way to make up our minds. Until, often as not and with a sigh of detectable relief, we revert to talk about 'nature' or 'God': rather than making our own decisions, we talk about letting 'nature take its course,' or issue desperate, prayerful pleas to God, thinking that things are surely out of our hands—it is instead due to God's will or blind nature, we believe—which only masks needed decisions. Many of these awful questions are insistently present as well for those of us actually involved in taking care of such patients, whether as physicians, nurses, chaplains, or ethics consultants—not merely, I mean to say, talking about patients with colleagues on committees, but rather with those who are actually facing a moment of extraordinary decision.

Beyond the exigencies of discussing ethical questions with those who actually have to face them, most of whom have never had to do this before, and perhaps helping them find some way to settle on some course of action—be they patients, doctors, nurses, family members, or any other—there is also, if we are honest, the arduous chore of putting that talk into written form at some point, into words that go beyond the moment, words that will truly get across the actual sense and feel of those disturbing situations. It is so very difficult, we then realize, to write without obscuring, concealing, masking, or even forgetting to mention precisely what was vital for others and ourselves as we then strived to understand what we were going through.

For me, James Agee said it best, in *Let Us Now Praise Famous Men*, that remarkable work for which he and Walker Evans were once commissioned, in order to write about unique individuals facing themselves across the awesome horizon of their own unique lives and deaths:

> For in the immediate world, everything is to be discerned, for him who can discern it, and centrally and simply, without either dissection into science, or digestion into art, but with the whole of consciousness, seeking to perceive it as it stands: so that the aspect of a street in sunlight can roar in the heart of itself as a symphony, perhaps as no symphony can: and all of consciousness is shifted from the imagined, the revisive, to the effort to perceive simply the cruel radiance of what is. (Agee and Evans 1939, p. 11)

But Agee was even more emphatic about the point I am trying to express with as much directness as I can. His first words about the project he was commissioned by the U. S. Department of Agriculture to write, along with Evans' extraordinary photographs—pictures and words about poor white dirt farmers in the South—give the compelling challenge to any who, like me, would dare talk with, much less go on to write about people in times of utmost distress. Writing about himself, Agee said:

> It seems to be curious, not to say obscene and thoroughly terrifying, that it could occur to [anyone]...to pry intimately into the lives of an undefended and appallingly damaged group of human beings, an ignorant and helpless rural family, for the purpose of parading the nakedness, disadvantage and humiliation of these lives before another group of human beings, in the name of science, of "honest journalism" (whatever that paradox may mean), of humanity, of social fearlessness, for money, and for a reputation for crusading and for unbias which, when skillfully enough qualified, is exchangeable at any bank for money and in politics, for votes, job patronage, abelincolnism, etc. (Agee and Evans 1939, p. 7)

Just this challenge has haunted my work; and, with Agee, I intend neither science nor art nor journalism: Is it ever possible to perceive anything simply "as it stands," especially while you are yourself there, too, *dans le milieu des choses*? And, supposing it can glimpsed, somehow: beyond the perceiving of things while standing in their midst, how then to talk with colleagues, of all sorts, much less,

later on, to write about those very things—without dissection, or analytic examination, neither absorption nor digestion, but straightforwardly, without clouds or
shadows or anything else obscuring—to say what must be said, and say it so that
you get it right? Can any of us "stand" in the face of that "cruel radiance of what is"
and tell it like it is?

How can we talk or write, when what we must talk and write about is so unique,
so singular in its immediacy and potency, and for that very reason seems, contrary
to everything we hope to do, so utterly unrecoverable? How put the unconditional
into words, how tell the unqualified uniqueness of individual people and their
actions, emotions, relationships, circumstances—more acutely, when they are so
deeply vulnerable and exposed? And how then go on to write about these very
encounters—standing then at still another remove from their ungetaroundable
immediacy—without obfuscating or distorting the very things that most mattered
while we were still enmeshed in the moment, in the circumstances that have left
such deep marks, still haunting us as we try to figure it all out...in writing?

8.5   Dwelling on the Unique and the Different

If you like others of us, are not content just to tell a story, but have to go on and
question why that is so, and the rest of what then quickly follows even a slight bent
toward philosophical wonder, things quickly become far more complicated.

To put myself on the block: what are the characteristics of an ethics consultant's
discourse, especially questions, such that it is seen—or perhaps better, is at all 'see-
able,' by patients, families, physicians, nurses, even oneself—as having to do with
ethics? Still more, what sort of thing is this listening and talking with people about
their lives and selves and circumstances, and then more especially writing about
those clinical encounters in which I am myself involved as a consultant, hence
being unavoidably affected and altered by what's going on even while also in our
very involvement affecting and altering what's going on?

The point I am trying to get clear about has, I think, to go by way of the device of
story-telling, which is at least one major reason for my having undertaken to do just
that on many occasions. For this device, I have come to believe, (Zaner 1996)
comes closest to embodying the significant characteristics of the 'talking and
listening' which I and any other 'ethics consultant' must centrally enact from the
very moment of being invited into and then entering someone else's life and
circumstances. I have long been centered on this point of emphasis. It was splen-
didly expressed by Paul Komesaroff:

> The major concerns expressed in the public debates about medical ethics ignore many of
> the most important issues. They ignore, for example, the finely textured and subtle nature of
> the interaction between doctor and patient and the social context in which it occurs. They
> ignore the manner in which problems are formulated within this relationship and the ways
> in which the various possible courses of action are identified. Most importantly, they ignore
> the delicate ongoing process of negotiation and compromise that characterizes human

relationships in general and in particular underlies any therapeutic interaction. (Komesaroff 1995, p. 66)

Or, in Agee's admirable bluntness: is it not "obscene and thoroughly terrifying...to pry intimately into the lives of an undefended and appallingly damaged group of human beings"? The question is humbling; I offer here only the following deliberations—along with, of course, my narratives (Zaner 2004, 2005a, b, 2007–2008, 2013–2014).

8.6 The Consultant's 'Involvement'

Two things should be acknowledged. First, there are important and difficult questions that stem from the various interactions of the people who first invite a consultant into their situation. On the other hand, there are the obstinate and awkward ethical issues that arise from the consultant's own involvement. To become involved ineluctably changes things, and not always for the better. This is true, both for those who issued the invitation, which has often been noted, and for the consultant, which has rarely been noted, perhaps because it is so awkward: I myself am then at issue and at risk, like it or not.

Note that people who engage in ethics consultation (in whatever manner—whether as professor with a student seeking advice, or as one of us working in a hospital) are not themselves the ones whose original situation occasioned the request for a consult about whatever questions it may be. When I become involved, it is at someone else's *invitation*—a point a consultant forgets at considerable risk; among others, obfuscating significant components of that very situation. Although not (at least initially[6]) of the consultant's own making, such situations as we may be invited into, for whatever reason, are *reflexive* in the multiple ways indicated.

By accepting such an invitation, the ethics consultant becomes part of (and thus he or she both influences and is unavoidably influenced by) what is and what is perceived (which are not always the same) as going on. Yet by virtue of that very acceptance of the invitation, the consultant at the same time bears and must accept the responsibility for whatever it is he or she eventually does and says as consultant —including what at some point the consultant may then write, and whatever the form that may take (from chart note to case report to, perhaps, a full analysis—or even narrative).

To become involved as an ethicist in a clinical encounter demands that one be strictly focused on the *individual* constituents (people, setting, circumstances, issues, etc.) *for their own sakes*. Such highly focused, individualizing concern is complex, as it is focused not only on the patient(s) and the veritable world every patient (including family, friends, values, concerns, occupations and preoccupations, etc.) presents; but

---

[6] As with all human action, things may change as the consultant begins and then proceeds; the consultant may well become a more central 'issue' than other ethical matters.

also on the professionals and their own respective spheres of individual and professional concerns, commitments, values, and the like.

The ethics consultant must be clinically attentive precisely and solely to the
individuals—any and all of them: patients, families, friends and acquaintances; as
well as the usual variety of professionals: physicians, consultants, nurses, social
workers, therapists, etc.—into whose situation he or she has been asked to become
involved. The consultant's concerns are, like any physician's, at one and the same
time *diagnostic* and *therapeutic*—to try and figure out what's going on and how
then one might be helpful in resolving whatever problems are eventually identified
and clarified.

The consultant thus seeks not only to understand but to be understanding: why a
request for the consultation came up in the first place, by or from whom it came,
much less what is actually learned at the scene about the variety of people,
interpretations and views each brings, etc. To help those whose situation it primarily is requires listening to and helping them figure out what they are facing and then
helping them to talk about it, to articulate their views in the attempt to help them
reach an understanding and, in light of that understanding, to identify and clearly
face those options, choices and associated decisions which, after as much discernment and deliberation as time and circumstances permit, seem most consonant with
who and what they are within their own life-context—including what they have
been and what they hope to become. Whatever else all that signifies, the ethics
consultant must be rigorously focused on just these individuals within their circumstances and with their concerns.

As with any such situation, there is inevitably a strong temptation to forget that
focus and instead become concerned with how you or I, the consultant, might see
things, and what you or I, the consultant, might decide to do, and the like. Which, of
course, is precisely what you and I should by no means ever do: it is not my
decision, not my life, not my values, and to the extent I am unable to put my own
concerns and values to the side, I have no business consulting with anybody else.

8.7   On Context and Clinical Deliberation: On Hide-Craft

It should surely be very clear by now that these are difficult issues, to say the least.
Of much greater interest to me than casting off still another opinion about one or
another 'case,' however, are several other, equally significant matters already
intimated. First, for all the disputes regarding clinical ethics consultation, there
are nonetheless certain characteristics that seem to me especially pertinent.

(1) Whatever the clinically presented problems may be, they are problems
strictly for the people whose situation it is—for instance, a specific family and
their physicians, but at some point any others who may be or become involved. Any
encounter presents its own set of issues, moral and other, and as noted already, these
are *context-specific* in the sense that working with and on behalf of such persons,
helping them appreciate and advising them regarding specific issues needing

resolution, and the like, requires a strict focus on what might be termed their own *situated understanding*, or each involved person's own *situational definition*.

To understand a clinical situation, there is nothing for it but to devise ways of allowing one to get at the concrete, situated ways through which the participants themselves experience and understand their situation, in whatever ways each endows various components (objects, people, things, relationships) with meaning, hence too how each one variously talks about what's going on—which may include, of course, at least some forms of writing about the situation (patient-chart notes, for instance). Having this in mind, I believe that stories, well-constructed and as faithfully rendered as possible, are precisely the 'device' called for. The 'logic,' if you will, pertaining to the uniqueness of human individuals is to be found in narrative language: stories capture what needs to be captured, precisely by virtue of their essentially 'indirect' ways of relating what needs to be related.

(2) To discover then try to understand what's going on in any clinical encounter—what troubles the people, what's on their minds, and thus to know what has to be addressed and how—requires cautious, attentive *listening* to and *probing* of their ongoing discourse as well as their conducts, the setting, and other matters that constitute this specific context. To 'say' or 'write' any of this, one must, I believe, learn the discipline of story telling and writing. And the stories of greatest if not sole relevance are those told by the unique individuals themselves—those whose situation it is—although, it is most often true, these stories are for the most part incomplete, at times poorly said, or may even be exercises in evasion.

(3) The ethics consultant is invited, and enters into an already ongoing clinical encounter between a patient (often, family, friends and others) and his or her physician (and others: nurses, medical consultants, social workers, and others). Every situational constituent, including any moral issue, is presented solely within ongoing relationships among patient and family and physician—at least, in its core form. And, the 'saying' and 'writing' of such complex affairs as these relationships, I believe, is best found in the telling of their stories.

The clinical ethics consultant, as I've emphasized, has an importantly different focus from either the physician or the patient, since he or she has to address the complex, ongoing *relationship itself*, attending to each of the integral constituents within that temporally unfolding contexture. The ethicist, then, bears the responsibility for enabling the pertinent stories to be told, and of equal importance, that need to be heard. The ethicist's focus must therefore be strictly on enabling people to listen to one another—which means, I think, to urge that people listen to the stories that are being told.

In a word, I should emphasize at this point that everything identified above constitutes a kind of discipline. It is, like medicine itself, an 'art,' or perhaps better and more accurately expressed it is a 'craft,' what might be termed *hide-craft*.

8.8   Writing About Clinical Situations

A certain light is thus shed on an intriguing issue buried squarely and in a way deliberately in my own efforts to tell the stories patients and others have tried to tell me.

Every attempt to write about clinical situations includes several quite different kinds of discourse. First, there is my sense of some situation as I earlier encountered it and now write down in my notes (both in the patient chart and in my own records). Second, there is my write-up of my impressions of the situation (often put into a note to the physician, or whoever else requested that I look into the situation). Then, third, there are the critical and interpretive commentaries about some of these situations, included in various lectures, professional articles, books and presentations given from time to time, along with the critical comments, oral and written, which other professionals make on my articles or presentations, and, of course, my responses to those comments. By the time you get to the latter, of course, you are quite a bit removed from the former—and even further from the original discussions, written and oral, that typically occur and are rarely if ever intended for eventual oral or written publication.

Beyond this, fourth, there are the stories I have written, using certain situations whose identifiable characteristics have been seriously altered and masked so as to respect the privacy of all concerned—still another remove from the original. Finally, fifth, there are the copious discussions, during and after the events related, with colleagues, both to help think through a situation thoroughly and to guide colleagues, students, and others to learn to think through such situations. All of these many times of listening, talking and writing may help and they may hinder, but all them figure in at least as essential components of the background to the narratives. The craft of story telling and writing thus emerges from what I've termed hide-craft.

This way of laying out the multiplicity of discourses, however, may too easily obscure a crucial point. We need to ask about the status of all these writings about some situation. Is any of the writing I've mentioned anything like a 'factual report,' a 'telling' of 'what actually occurred'?

If so, then the proper response to anyone's concern about 'the facts' should be an answer to the effect that, all things considered, what was given is, to the best of my recollection and reconstruction, what 'really and truly' occurred. What else could the question possibly mean? Perhaps I've invented a story, or certain parts of it—to protect the innocent, I might say—even while I've tried to make it 'real' and make it look 'factual.'

Writing as factual reporting implies that the author is authoritative: the information provided is judged by the author to be all that is needed for an appropriate understanding of what went on. And, since the author is presumed to be the 'expert,' then any concerns about 'what the facts are' can be answered only by direct appeal to what has been factually reported. Since, however, what is reported about is unrecoverable, the moments of its occurring past, there is simply no way to determine whether what was reported is what 'in fact' actually occurred. We have only the reporter's words and his or her 'word' about what is reported. Trust is as vital here as it surely is in the clinical event.

Unfortunately, if writing a 'clinical report' is only 'factual recounting' in this sense, it is then supposed that the (often sole) person who could possibly know what 'actually (really and truly) happened' is the very one who also now makes that very claim. The (often sole) source of 'information' about the 'case' is and can in many instances only be the one who did the consulting and then reported on it—and who is accordingly also the only person capable of settling disputes, correcting mis-understandings, drawing appropriate conclusions, making pertinent recommenda-tions, etc. A wonderful example of conflict of interest!

If, that is to say, writing about such situations is just a matter of factual reportage—which may well imply a notion of 'the reasonable ethicist' or 'ethics expert' comparable to long-standing legal idea in medicine of 'the reasonable physician'—one underlying presumption resembles something on the order of the claim that people who are not directly involved cannot possibly know what's going on in a specific situation unless similarly involved in it—and probably not even then, since to be an 'expert' is of course also to know what exactly to look for, how to report it all, and the only 'ethics expert' is of course. . .there simply is no way out of that question-begging circle.

There are all sorts of other perhaps more obvious, if also serious problems with that too-common approach to writing about clinical situations. Perhaps most per-tinent here, however, is that such a conception of writing ignores, and must ignore, one of the most prominent features of every clinical situation: that each is packed with interpretations, each type of which is presented as in need of a sort of unpacking. To get at this, permit a brief excursus.

8.9   An Historical Excursus

In the ancient world, especially in the skeptical or 'Methodist' tradition—whose methodical views derived strictly from the individual's own clinical experience (Edelstein 1967, pp. 193–99)—it was believed that each illness or injury was unique precisely because every person who fell ill or was injured was unique and reacted differently. Symptoms were taken as signs of the body's own powers (*physies*) called forth to combat the influences of bad living, noxious environment, or both. As is stated in several of the Hippocratic texts, "the *physies* are the physicians of disease" (*Epidemics*, VI), and the clinical healer is their servant, acting to support these powers (*Epidemics*, I). Or, as Eric Cassell says, "the illness the patient brings to the physician arises from the interaction between the biological entity that is the disease and the person of the patient, all occurring within a specific context," and as these differ so must the physician's responses differ—even in the case of the same patient coming down with 'the same' disease at different times (Cassell 1985, pp. 4–5).

Moreover, Arthur Kleinman has suggested that, "When we speak of illness, we must include the patient's judgments about how best to cope with the distress and with the practical problems in daily living it creates" (Kleinman 1988, p. 4). For this, he says, it is necessary to utilize one's own "common-sense" in categorizing and explaining the kinds of distress (moral, social, etc.) brought on by pathophysiological processes. In different terms, utilizing the interpretive categories ingredient to everyday life, it is clear that each of us "order [our] experience of illness—what it means to [us] and to significant others—as personal narratives" (Kleinman 1988, p. 49). Frequently, however, neither the sick person nor his or her family is able to express the full story, the illness experience, adequately or accurately—surely a requirement for judging whether s/he is truly informed, uncoerced, and capable of making decisions.

Among other things, as one patient plaintively remarked, while "you want your doctor to understand," we tend to be "too timid" around them. Another patient also pointed out how important it is to talk, to "communicate;" however, he continued, "there are lots of feelings that are hard to put into words, especially if you've never had the feeling before."

Patients (to write only about them for the moment) experience their ailments, talk about them, and interpret them. A core clinical task, in Kleinman's words, is the "empathetic interpretation of a life story that makes over the illness into the subject matter of a biography... [which] highlights core life themes—for example, injustice, courage, personal victory against the odds" (Kleinman 1988, p. 49). For this interpretation (of the patient's own interpreted and expressed experience), it is necessary to piece together the patient's telling of what has happened as it is embedded in the patient's complaints and talk about them. This 'piecing-together' is, of course, another work of the hide-craft of interpretation developed with the patient, family and health professionals, and eventually also expressed in written form.

Beyond this 'reading' of the patient's experience and telling of his or her experience, of course, every clinician (whether physician, nurse or ethics consultant) experiences the patient (smells, sees, touches, listens, etc.), and engages in the diagnostic work that must also be done: physical examination, tests, measurements, instrumental visualizations, and the like, all conducted so as to determine with as much precision as possible what's happened to the patient and what types of available therapeutic alternatives there might be. These, too, are works of interpretation—often conveyed in different ways to patients—which thus become components in the way patients come to interpret and reinterpret their conditions (diagnoses) and futures (prognoses). Thus are interpretations mixed in with other interpretations, becoming factors in subsequent re-interpretations, and so on. Expressed differently, our experiences are *thoroughly storied*, even while most of the time they are only partially told.

The significance of these multiple and complex interpretations is in part the need to develop the linguistic skills—principally, listening and learning to talk about sensitive topics with candor and completeness—that are necessary to understand what's going on in a patient's life, what's implicit in his or her discourse (questions, responses, etc.). Such a complex interpretive discipline, hide-craft, is or ought to be quite as important as any of those commonly associated with physical diagnosis.

8.10 Clinical Interpreting

I can now say more clearly what I've been driving at: while the kind of clinical conversational attunement that is focused on patient experience and self-interpretation has only begun to be more appreciated in the health care professions, especially medicine, precisely this methodical craft or discipline is *the central feature of clinical ethics consultation.*[7]

In different terms, to be oriented, understanding and sensitive in any clinical involvement, ethicists must learn to be as precise and careful in attending and listening as the physician who auscultates a heart or palpates a spleen (Cassell 1985, p. 4). Physicians, in Kleinman's interesting but, I think, misguided, analogy, are "naive realists, like Dashiell Hammett's Sam Spade, who are led to believe that symptoms are clues to disease, evidence of a 'natural' process, a physical entity to be discovered or uncovered"—incorporating a positive tendency to "regard with suspicion patients' illness narratives and causal beliefs" (Kleinman 1988, p. 17).

If I am correct in the way I have conceived the work of the clinical ethics consultant, unlike physicians (focused on each patient) and unlike patients (focused on their own condition and prospects), the clinical ethics consultant must be *wholly unlike* any "naive realist:" like a more philosophical Sam Spade, perhaps, ethicists must bring a clear-headed, strong reflective presence to clinical situations, to ensure that the fundamental questions of moral worth are not avoided, but are instead at the very center of every clinical conversation and decision (Zaner 2013–2014).

The patient and family, but also the physician and nurse, and others who may become involved and influence the course of a patient's condition, thus exhibit interpretive methods quite as much as any one else. They are, in Kleinman's words, like "revisionist historians," "archivists," "diarists," even "cartographers," who constantly search their pasts for present meaning, record the most minute difficulties on the map of changing terrain of ongoing illness, and focus on the "artifacts of disease (color of sputum, softness of stool, intensity of knee pain, size and form of skin lesions)" (Cassell 1985, p. 48).

Ethicists, I am urging, are hunters and gatherers at the same time, in any case they are listeners and collectors of the almost always-partial stories that make up any and every clinical encounter. Beyond hunting for and gathering, collecting, such stories, they are also witnesses and guarantors, ensuring that every clinical narrative has its chance to be told and receives its appropriate hearing, that every 'voice' has its chance to be heard.

---

[7] A good case might be made for conceiving clinical ethics as such an interpretive discipline, as I've argued elsewhere.

8.11 Concluding Reflections

I must not ignore what I believe is a crucially important feature of writing, especially narrative writing: in a word it is a *way to discover* and understand the sense of what was going on in the initial situation now being told. Consider only those moments of initial writing when we are trying to get our thoughts clear, trying to find adequate and accurate ways of expressing what went on in a particular clinical encounter. Those initial, always laborious moments of writing seem most of all a kind of discovery, when we try out first this or that expression, phrase, or word, to listen to whether it says rightly what is on our minds regarding this or that moment in the clinical conversation—that we've 'had it right.' To appreciate that such phases of writing are, in truth, moments of genuine discovery, is to acknowledge as well that even that much-sought-for final version, the one designed for full public display in some published form, may itself continue to be a kind of discovery.

If this is granted, then the status of the writer must be thoroughly reconsidered: rather than simple 'data-gatherer,' much less mere 'recorder' of facts—and their adjudicator in case of dispute—the sense of writing-as-discovery suggests that the writer is more hunter than gatherer and has the orientation more of inquirer than recorder, more interrogator than settler of disputes, more the posture of one still learning than one of having-already learned. It is the *craft* of listening for and careful attending-to what is often unspoken, unnoticed, and thus unheeded by those whose situation it is most of all.

I want to emphasize, too, that every moment of consulting must, as I understand these matters, be shared—impressions discussed, initial judgments tested, implications explored, the 'lay of the land' properly told. Conversations and writings need to be continually submitted to others, for their understanding but also for their critiques. Understood in these terms, writing may be one aspect of a more general method. Pursuing this notion, we might then note that the initial piece of writing often has the form of a narrative, albeit tentative and partial.

So far as moral issues are in the strictest sense context-specific and -dependent, and to the extent that clinical encounters are invariably a form of passionate drama focused on and by what matters most to those whose situation it is, it then makes good sense to conceive case write-ups as having narrative form, even if only nascent. They are, if you will, somewhat like stories, or at least anecdotes, waiting to be told. Rather than 'factual reports,' they are dramatic scenarios that evoke the whole array of emotional, volitional, and valuational themes and transactions so characteristic of clinical encounters, and human life more broadly.

My 'way' of writing has evolved, for better or worse, into this: a perhaps untidy mixing of straightforward story telling with occasional reflections. I have done it this way because, put most succinctly, I must: that is what happens in these encounters, by everyone involved and not merely by me. People not only act and interact; we also think about it and then talk about that, and ask each other is this or that was what happened or was best expressed; and as our actions are only sometimes 'right on,' so for our thinking—and, therefore, our telling of the story, our trying to get it right. So, I find myself mixing up what otherwise may seem very different ways of writing—factual reporting, narrative relating, reflective evoking—but I do so because it is all intimately part of what must be told. "That," Rosenblatt once keenly observed, "it is what we were meant to do." Indeed, I am constantly seeking that "monumentally elusive tale" which every clinical encounter evokes.

If that is true, then writing about my encounters can only be a voyage of continual discovery. With the beacon of Agee's example in front of me—for all his lamenting, he did after all go on to write, at incredible length and masterfully, even if he was prying "intimately into the lives of an undefended and appallingly damaged group of human beings"—I too find myself having nonetheless to enact my own sort of prying, hoping that, a little like Agee, I can gradually discern each of the situations I write and have written about, "centrally and simply, without either dissection into science, or digestion into art, but with the whole of consciousness, seeking to perceive it as it stands," so that I may, perchance, "perceive simply the cruel radiance of what is"—the disciplined craft of telling, I believe, can only be a story, mixed up, however painfully, with continual reflections.

References

Agee, James, and Walker Evans. 1939. *Let us now praise famous men*. New York: Houghton Mifflin.

Blythe, Ronald. 1991. Introduction. In *The death of Ivan Ilyich*, ed. Leo Tolstoy. New York: Bantam Books.

Cassell, Eric J. 1973. Making and escaping moral decisions. *The Hastings Center Report* 1: 53–62.

Cassell, Eric J. II 1985. *Talking with patients*, vol. II. Boston: MIT Press.

Edelstein, Ludwig. 1967. *Ancient medicine*. Baltimore: John Hopkins University Press.

Kierkegaard, S. 1954. *The sickness unto death* (with *Fear and trembling*). Princeton: Princeton University Press.

Kleinman, Arthur. 1988. *The illness narratives*. New York: Basic Books, Inc.

Komesaroff, Paul. 1995. From bioethics to microethics: Ethical debate and clinical medicine. In *Troubled bodies: Critical perspectives on postmodernism, medical ethics, and the body*, ed. Paul A. Komesaroff. Durham/London: Duke University Press.

Pellegrino, Edmund D., and David C. Thomasma. 1981. *A philosophical basis of medical practice*. New York/Oxford: Oxford University Press.

Schutz, Alfred, and Thomas Luckmann. I: 1973. *Structures of the life-world*, 2 vols. Evanston: Northwestern University Press.

Schutz, Alfred, and Thomas Luckmann. II: 1989. *Structures of the life-world*, 2 vols. Evanston: Northwestern University Press.

Zaner, R.M. 1981. *The context of self*. Athens: Ohio University Press.

Zaner, R.M. 1996. Listening or telling? Thoughts on responsibility in clinical ethics consultation. *Theoretical Medicine* 17(3): 255–277.

Zaner, R.M. 2004. *Conversations on the edge: Narratives of ethics and illness*. Washington, DC: Georgetown University Press.

Zaner, R.M. 2005a. But how can we choose? *The Journal of Clinical Ethics* 16(Fall): 218–222.

Zaner, R.M. 2005b. Parental voices: Randal Lewis Morris was born. In *Parental voices in maternal-fetal surgery*, a special issue of Clinical Obstetrics and Gynecology, vol. 48, no.

3, ed. Larry R. Churchill and Mark J. Bliton, 548–561. Philadelphia: Lippincott Williams & Wilkins.

Zaner, R.M. 2007–2008. "It must have rained hard that night in March," a story. *Langdon Review of the Arts in Texas* 4: 210–222.

Zaner, R.M. 2012. *At play in the field of possibles: An essay on free-phantasy method and the foundation of self.* Bucharest: Zeta Books.

Zaner, R.M. 2013–2014. "The indomitable Rachel Bittman," a story. *Langdon Review of the Arts in Texas* 10: 68–79.

Chapter 9
Visions and Re-visions: Life and the Accident of Birth

9.1 A Preliminary Observation

Much has happened in the years since I wrote much of the previous Chapters, especially in the world of health care, medicine in particular but even more in medical and bio-medical research. The latter, indeed, is substantially responsible for many of the significant changes in clinical practice, diagnosis and prognosis in recent times. On reflection, it remains somewhat unclear to me that these changes, such as they may be, will also alter the moral themes and basic approach of the preceding Chapters. But since so much has in fact happened, it seemed to me only appropriate to include the following reflections on what had come to be known early in the past two decades as the 'new genetics' (Zaner 2005, pp. 177–207).

9.2 Of Fiction and Fact in Science

In the late 1950s, the English science fiction writer, James Blish, wrote a charming little novel suggestively entitled, *The Seedling Stars and Galactic Cluster* (Blish 1957). It had a simple premise, as inventive as it was remarkable for its prescience. Habitable planets for human beings had become premium, for straightforward reasons. Interstellar travel had become routine even as the population had long since burgeoned beyond Earth's and other planets' resources. The planets discovered, however, mostly turned out to be fiercely uninhabitable. Making them habitable required immensely complicated, expensive and only rarely effective labor, by

This Chapter, first developed as a presentation for the University of Scranton conference, 'Genetic Engineering and the Future of Human Nature,' April 6–8, 2001, was later revised and published. The present version is a revision of those papers.

© Springer International Publishing Switzerland 2015
R.M. Zaner, *A Critical Examination of Ethics in Health Care and Biomedical Research*, International Library of Ethics, Law, and the New Medicine 60, DOI 10.1007/978-3-319-18332-9_9

means of a process Blish called "terra-forming." To make a place human-friendly, in these terms, required either transforming that environs and atmosphere, or protecting people from its hazards by special shelters, breathing apparatuses, and the like.

But the science of biology, Blish also postulated, had already undergone a sweeping revolution—the beginnings of which were already apparent when his novel first appeared and, we have since become acutely aware, continues to undergo its consequent transformations, matching if not surpassing the earlier one in physics. In the novel, biological manipulations are routinely done and increasingly designed for population projects focused on the most elementary reproductive life-processes, including cloning and other types of genetic engineering; along the way, medicine had been transformed as well in every respect.

Blish's tale is delightful. In his imaginative hands, the deliberate, literal re-designing of human individuals by other human individuals is an accomplished fact, bringing about changes which need neither centuries of gradual evolutionary change nor spontaneous mutation, only the ingenuity and sportive inventiveness of highly powerful biomedical scientists possessing 'the secret of life.' Such scientists are now avidly in pursuit of ever-new ways to redesign people, especially to enable the settling of populations in otherwise hostile environments—not to mention the extreme hardships of space travel.

Notably, much the same aim was overtly advocated in the 1960s by the Nobel Laureate in genetics, Joshua Lederberg, (cited in Kass 1997, p. 17) and in the late 1970s by two Nobel Laureates, neuroscientist Sir John Eccles (1979) and prominent immunologist Sir Macfarlane Burnett (1978) and first became a reality for complex animal vertebrates in the 1990s by Ian Wilmut and colleagues at the Roslyn Institute in Scotland (Wilmut et al. 1997). What Blish only imagined is already more than a mere promissory note, to the delight or dismay of a host of commentators.

In the early 1990s, another Nobel geneticist, Walter Gilbert, expressly, if with some hyperbole, portrayed just that underlying vision as the "holy grail" of our times. The secret foundations of human life (in the multiple shapes proteins can take) seem now to have come very close within sight. The unraveling, mapping and sequencing of the human genome accomplished in countless projects around the world, Gilbert averred, promises to "put together a sequence that represents. . .the underlying human structure. . .our common humanity." Soon, he was convinced, we'll be able "to pull a CD out of one's pocket and say, 'Here is a human being; it's me'!" (Gilbert 1992, p. 95).

Thanks to that, the response to the ultimate questions of human life was for such scientists already well on the way to being definitively answered by molecular sciences, especially genetics. What is distinctively 'human' either is a matter of genes, or is in the genes, and thus not to be found in the quaint metaphysical quests that moved Plato or Aristotle, Aquinas or Occam, Kant or Heidegger. Something like full circle will then be reached, for at the time of DNA's discovery shortly after the publication of Blish's fascinating novel—what a 1961 *Life* magazine cover declared as the "secret of life," and which Kurt Vonnegut soon after satirized in his classic novel, *Cat's Cradle* (1963). It was then obvious to many (Eccles 1970) that the new genetics was indeed "the holy grail" of science and society. The human genome soon came to be regarded as the secret hiding place of self, indeed of life itself—a notion already perhaps somewhat passé, as there quickly began to be talk of digitizing the entire genome onto ever tinier chips that could then be embedded in any cell—a sort of postmodern covert mole always on call and ready to be pulled out, read and, if need be, cloned.

This motif is historically fascinating as well, for it is of a piece with one of the core convictions in medicine's long history, as articulated in one or both of two fundamental visions. Ancient physicians were struck by the ways in which the human body and soul could be changed by dietary regimens. The prominent Greek physician in the Roman empire, Claudius Galen, went so far as to assert the need to "clear the path for using bodily factors to elevate man beyond the possibilities of purely moral teaching" (Tempkin 1973, p. 85). Galen's concern was as much if not more to improve the human condition as it was to treat diseases—and in this was close to Blish's biologists in his vision. Contrary to Galen's apparent aims, however, reports from the Genome Project mainly highlight the therapeutic potential of new discoveries while almost always downplaying any such eugenic designs that captivated both the ancients and much of contemporary science fiction.

Walter Gilbert's colorful way of portraying that visionary theme to the side, it is important to point out that he is hardly alone. Other distinguished scientists, as mentioned, were equally caught up in and have long articulated portions of that genetics vision. Nevertheless, despite the hype and repeated promises of therapy that are invariably part of the frequent announcements about new genetics' discoveries, anyone seriously considering these and related issues is well-advised to bear prominently in mind that "haunting memory—that most of the world still consists of things and creatures that neither scientists nor social theorists had any hand in making" (Winner 2005).[1] That, and the apparent need to be reminded of the not always agreeable record of some technological projects—one need not go much beyond that of nuclear power to fuel our modern age's incessant appetite for the always more and ever new, while pretty much ignoring and evading essential questions such as the disposal of the inevitable toxic wastes. Not to mention the uneasiness we feel when attending carefully to the record of disasters and abuses that is so much a part of the recent history of bio-medical research—the radiation experiments first publicly disclosed by Representative Edward J. Markey (D., Mass) in 1986,[2] the syphilis experiments at Tuskegee, (Jones 1981) or the many questionable ones highlighted by Beecher (1959, 1966) and others (Howard-Jones 1982; Curran 1982; Annas and Grodin 1992).

How can we make sense of this? An indirection will be useful.

---

[1] Langdon Winner, "Resistance is Futile: the Posthuman Condition and Its Advocates." In Harold W. Baillie and Timothy K. Casey (Eds.). *Is Human Nature Obsolete? Genetics, Bioengineering, and the Future of the Human Condition*. Boston, MA: MIT Press, 2004, pp. 385–310.

[2] In 1986 congressman Markey released records detailing experiments by the U. S. government between 1940 and 1971.

9.3   Mendel's Dwarf

In Simon Mawer's engaging novel, *Mendel's Dwarf*, (Mawer 1998)[3] the principal character is Dr. Benedict Lambert, who is a world-renowned geneticist, the great-great-great nephew of the justly famous monk and geneticist, Gregor Mendel, and a dwarf (achondroplasia).[4] Whether as a professional delivering a scientific lecture, a student sitting in a college class, or simply as a citizen walking the streets of a city, Ben is said to be made acutely aware of himself by phenotypically normal people who invariably gawk at him, no matter where he happens to be.

After his much-anticipated lecture that opens Mawer's novel—invited by the Mendel Symposium at the Masaryk University of Brno to celebrate the life and work of his great-great-great uncle Gregor—the secretary of the association greets Ben and, this "large and quivering mountain of concerned flesh," as Ben calls him, says, "Gee, Ben, that's wonderful. So brave, so brave...." (p. 5) At which Ben thinks to himself: "Brave. That was the word of the moment. But I'd told Jean [his lady-love] often enough. In order to be brave, you've got to have a choice." (p. 5) And, of course, choice about his dwarfism was what he never had. Rather, like the rest of us, he had only that "tyranny of chance" when just one of the countless spermatozoa from his father's erupting orgasm was subtly and successfully attracted to, then penetrated his mother's ovum and, shedding its tail, managed to impregnate and fertilize—those magic moments of entrance, penetrance, implantation and conception thanks to which a specific child, Ben, is conceived, borne by his mother, and later born into the world.

Later, when the headmaster of his elementary school remarks, after Ben suffers a typical round of teasing from his classmates ("Mendel, Mendel, Mendel's dwarf"), "it's a problem you have to live with." He then objects silently to himself: achondroplasia is not like premature baldness, a birthmark, or a stutter, "it is me. There is no other" (p. 21).

At one point, Mawer has Ben excitedly announce how he loves Dinah—the first girl he's ever kissed (rather: "*she* kissed *me!*")—after he helped her get through a genetics class. Declaring his love, and, despite his help, she dismisses him with a flip, "thanks everso," and a final "it can't be." He then replies, "I'll say it for you: you can't love me because I'm hideous and deformed, a freak of nature, and people would stare...You can say this: 'I would love you if you weren't a shrunken monster'" (p. 52).

---

[3] Note: all citations in the text are from this edition of Mawer's novel.

[4] Among other features, such dwarfs exhibit a large skull, with a narrow foramen magnum, and relatively small skull base. The vertebral bodies are short and flattened with relatively large intervertebral disk height, and there is congenitally narrowed spinal canal. It is caused by a change in the DNA for fibroblast growth factor receptor 3 (FGFR3), which causes an abnormality of cartilage formation and such dwarfs are thus of short stature. The cause is either a sporadic mutation or is an autosomal dominant disorder.

He goes on later in his life to have a highly successful career in human genetics, and was invited to be the Mendel Society's honored speaker at the Brno Conference, not because of his kinship with Father Mendel, but rather because he had identified the achondroplasia gene, the very gene whose flawed working (or whose correct working with an incorrectly 'spelled' gene) resulted in the dwarf, Ben. When his results became public, the media took a shine to him; a major newspaper reported on the discovery, with the headline: "Dwarf Biologist Discovers Himself." His sister telephoned him to tell him about it, reading the text of the report to him: "Super geneticist Ben Lambert has finished his search of a lifetime. Genetic engineering techniques and years of patience have finally led him to discover the gene that has ruled his own existence, for Ben, thirty-eight and a researcher at one of the world's leading genetics laboratories, is…a dwarf. Little in body but big in spirit….," as the new media darling reported (pp. 242–3).

It is clearly awkward at best to contemplate Ben's condition from the perspective of medicine in its traditional emphasis on restoring body functions and organic processes lost by illness or injury, or compromised by congenital or genetic condition. In the first place, though severely compromised by being a dwarf, he isn't sick in any conventional sense operative in this tradition. Even while shunned in multiple ways by other people, he is also a genius—and in this respect, he enjoys a privilege of place and the admiration of other people and of colleagues, especially among those in restorative medicine (Zaner 2001a, b). Despite that, as a dwarf he is *beyond* the limits of restorative medicine, *outside* its purview, unless he is sick in a conventional sense (flu, pneumonia, cancer, etc.). If the dwarf is outside the conventional and the customary, and if clinical, restorative medicine can do nothing for his condition as a dwarf, what exactly *is* he in conventional terms?

As he knows intimately, being outside the usual and the routine means that he is phenotypically *ab*normal—despite its having resulted from the "tyranny of chance" of disfiguring achondroplasia (though, of course, we are all so configured, if not abnormal and dis-figured, by chance's tyranny). As Mawer remarks about Ben, he may be "hideous, deformed, a freak of nature…a shrunken monster," but he is neither "sick" nor "injured." In this sense, geeks and freaks, dwarfs and hybrids, and other genetically or congenitally disfigured individuals, are socially constructed by 'phenotypically normal' people as beyond repair and thus fit mainly for carnivals and back-street sideshows. Medicine's restorative approach to illness and injury cannot bring such freaks and hybrids back even approximately to accepted norms, social or other. A dwarf may be puckish, an imp, or a good fellow, while another may be a rogue and a cad, but all of them are beyond the social limits due to the "tyranny of chance" of their births and how "the others" construe that.[5]

As I noted, Ben *can* be restoratively treated: if he gets the flue, renal disease, cancer, or any of the many diseases which can afflict any human being, he is then

---

[5] Another novel well worth taking quite as seriously as Mawer's is Katherine Dunn (1983). Dunn lays out precisely these variations of anomaly, personality and values among the children deliberately conceived by their parents to be freaks.

for the most part just like any of us normals. But when he is not conventionally ill the reasonable and even required thing for the restorative physician to do about Ben is to stand back from him, trying (most often unsuccessfully, as Ben learns) not to judge his condition as abnormal—precisely *his* normal condition.

Yet, if you were to ask him, Ben certainly does try, at times desperately, to be like others—if only he could do that. Faced with Ben in whatever situation, restorative physicians must surely sense his plight and, sensing it, would surely wish it were otherwise. The point is obvious when Ben talks with Dinah, or later with Jean. His body is seen as lacking (although it is clear that he performs sexually quite well), by others and thus by him. Hence, *he* is lacking, Ben is less than he should be, and just this targets him as the object of gawks, the butt of jokes, a creature of side-glances and sly pranks, ridiculed, ignored, abandoned, by-passed, looked-over, mocked.

Ben would obviously have it otherwise. In fact, this becomes evident when, having become expert in *in vitro* fertilization and embryo transfer (IVF/ET), he agrees to perform the procedure for Jean using his own sperm—and suffers the choices with which, as a dwarf with its genetic roots, he is then confronted. Yet, as his training in genetics makes clear as well, this signal event in the novel finds him—and us, the readers—at a very different place than we might have expected. For now even IVF/ET is transformed when it is in the hands of a geneticist accomplished in the arcane arts of recombinant DNA techniques—and the sperm donor for the process. Now, truly awesome issues, previously only barely beneath the surface, explode onto the scene. More on this in a moment; for now, other aspects of the phenomenon need to be probed.

9.4   The 'Scandal' in Medicine's 'New Paradigm'

Most of us sense the frustration of being unable to do anything to change things for someone like Ben. We sense as well the injustice in our social values that work so powerfully and severely to circumscribe his life. And there is a cutting irony: Ben is himself a renowned geneticist, the descendant of Gregor Mendel—also a geneticist—and Ben has succeeded in identifying the 'dwarf gene.' Indeed, using rDNA techniques, Ben is even capable of splicing the achondroplasia gene out of his and Jean's resulting embryo—or, as happens in the novel, of choosing to implant either that, or an unaffected embryo into Jean's uterus.

Ben thus knows well the sharp edges of the new genetics. After delivering his speech to the Mendel Society, he meanders through the tiny village where his great-great-great uncle had worked, and reflects: "This acre of space was where it all started, where the stubborn friar lit a fuse that burned unnoticed for thirty-five years until they discovered his work in 1900 and the bomb finally exploded. The explosion is going on still. It engulfed me from the moment of my conception. Perhaps it will engulf us all eventually" (p. 10).

Although a science of genetics could not truly get going until Mendel's work had been discovered and understood, this has now been done; reality has swiftly caught up with Blish's and Mawer's imaginative skills. The human genome has now been completely mapped and sequenced—and although understanding lags far behind, it too is picking up momentum. But just here something quite different has appeared. Now, unlike any time in medicine's prior history, the ground has shifted and what is still called 'medicine' might well soon be capable of *doing something* even for an individual like Ben, will be capable of what could hitherto only be barely imagined. A fundamental limit in restorative medicine seems now more a challenge and problem to be surmounted by molecular medicine: the vision in Blish's novel has flowed into that in Mawer's.

To be sure, there is still a kind of limit: it remains true that nothing can be done at the moment to change Ben's body into a phenotypically normal one. What's already happened cannot be altered—at least in his case, at least not yet. In other cases (cystic fibrosis, breast cancer, and others), the same techniques Ben uses to discover his gene and later uses for his and Jean's embryos, can now be used with very different aims in mind—even, it may be, for the fully formed child or adult. At least, that is part of the promissory note of the unraveling of the genome, the location and functional identification of each gene.

The implications of this are remarkable. Rather than being beyond the limit or norm, much of the sort of human affliction hitherto outside medicine seems now capable of being brought inside. I mean: not even the gnarled body of an achondroplastic dwarf, in the end, is any longer—as within traditional restorative or curative medicine it had to be—thought to be beyond the pale, no more than, say, is the neural regeneration of a quadriplegic's spine. Where the traditional view of medicine put in place the long-standing, still-viable endeavor of restoration, that approach and its limitations are now being challenged and potentially changed, decisively. Beneath the awesome potency that haunts the phenomenon of human cloning lies this astounding possibility, this fundamental shift in what medicine, disease, and health have long been thought to be all about.

This is not just unparalleled but may seem appalling. Thinking about just these matters, the science historian, Hans-Jörg Rheinberger[6] concludes that a "new medical paradigm: molecular medicine," (Rheinberger 1995)[7] already ongoing for the past century, has more fully blossomed over the past several decades and is well on its apparently unstoppable way to take over the entire garden. He insists, however, that there is a fundamental scandal at the core of this new paradigm, very much like what Claude Lévi-Straus diagnosed as the core of the incest taboo (Lévi-Straus 1967, p. 10). Rheinberger notes (p. 258) Jacques Derrida's (1978, pp. 278–93) observation that this taboo is right at the edges of, if not actually within the "domain of the unthinkable," for it challenges the very thing that makes possible the

---

[6] A microbiologist, Rheinberger is also Director of the Max Planck Institute for the History of Science in Berlin.

[7] Subsequent citations from this article are cited textually.

distinction and opposition between "nature" and "culture." That distinction, Derrida had contended, has for centuries been at the heart of philosophy and theoretical thinking generally. Thus, the very possibility of philosophical conceptualization itself has come under severe threat, if not actual collapse, as that distinction loses its sense in the presence of this scandal.

Lévi-Straus had argued, "everything that is universal in man belongs to the order of nature and is characterized by spontaneity, and that everything bond to norms belongs to culture and is...relative and...particular." From this, he then identified the epitome of scandal, "the incest prohibition," which, he thought, "escapes any norm that...distinguishes between... culture and nature. It leaves in the realm of the unthinkable what has made it possible" (p. 258).

Although he has far more subtle issues in view that I must ignore here, Derrida emphasizes that the taboo exists solely within a context that accepted the opposition between nature and culture. The fact is, he says, the scandal is "something which no longer tolerates the nature/culture opposition he has accepted," (p. 283) and thus renders "unthinkable" what was earlier believed to be "thinkable"—and thereby both philosophy and science become deeply problematic.

In just the way that the incest taboo is scandalous, Rheinberger is convinced there is also a scandal at the heart of the "new medicine." We may catch a glimpse of what he has in mind if we think about a key feature of biomedical science when conducting research involving human subjects: informed consent. If medicine's very point is to help sick, compromised people who cannot help themselves—people who for those very reasons are multiply disadvantaged and at their most vulnerable—how could there ever be any question about informing people and ensuring that nobody takes advantage of them? Why require what on the other hand seems perfectly obvious? Yet, just that doctrine has become a centerpiece of medicine and biomedicine—not only in research but in daily clinical practice as well.

In both cases, there would be no need either for a taboo (in the case of incest), or the legal requirement for informed consent (in the case of human subjects research). If vulnerable patient-subjects were not abused in some manner in the first place, the demand to obtain informed consent would be pointless—as would a taboo on incest, if no parent or sibling engaged in sexual activity with child or sibling. Just as incest seems but barely capable of being spoken or even thought, so is it scandalous that otherwise decent people who are researchers (not simply those who were Nazis) must be subject to the rule of informed consent—as if they could not be trusted. Distrust seems, indeed, to precede either taboo or human research.

Rheinberger is in any event very clear about what he think is as scandalous as the incest taboo:

> With the acceleration of a historical, irreversible alteration of the earth's surface and atmosphere, which is taking place within the span of an individual human's lifetime; with the realization that our mankindly, science-guided actions result, on a scale of natural history, in the mass extinction of species, in a global climatic change, and in gene technology that has the potential to change our genetic constitution, a fundamental

alteration in the representation of nature is taking place, which we are still barely realizing. (p. 260)

To be sure, therapeutic discovery and diagnosis continue to occupy the limelight of human genetics research—even with its newly acquired name, genomics—with actual treatments and understanding lagging behind.[8] Nonetheless, the regularly used discourse about (and often, presumable justifications for) genetics projects, and probable future reality of genomics, is that clinical practice will be totally transformed as new genetic knowledge leads eventually to effective treatment modalities. With that eventuality, a wholly new meaning of 'health' must shortly follow: more a matter of healthy genes (with the ability to keep them healthy) than of the absence of health or the workings of some pathological process or entity.

At this point, it seems to me necessary to take a few cautious steps of my own into the seemingly unforgiving unthinkable.

9.5 Beginning to Think About the Unthinkable

(a) In traditional, restorative medicine, there is nothing that can be done for Ben's condition. If he is injured or becomes ill, of course, as much can be done for him as for any other—taking into account that his condition may itself require one or another regimen. While changes of social attitudes and acceptance, along with support to pursue accepted goals or careers, even if not done or not done well by those who meet or know such dwarfs, can be recommended, they are plainly sufficiently rare as to prompt some cynicism.

But is this sort of encouragement even medicine's business? Should physicians be involved with or even concerned about the mistreatment Ben regularly receives? Doesn't this sort of thing fall to others—social workers, ministers, rabbis, or therapists? In the end, why should any of us be much concerned about dwarfs like Ben? After all, what we were born with is neither more nor less thanks to chance than is Ben's condition. For that matter and unlike most of us, Ben is a famous scientist who achieves an appointment to a famous institute. What need does he have for anything from medicine or the rest of us? If he is singled out for special consideration, doesn't this simply defeat the very purpose for special consideration?

Still, even considered merely as a body, we are obliged to recognize that while currently nothing can be done for Ben and others, yet in the new genetic, molecular model, such people may no longer be so obviously off the medical agenda or research, and in any event their progeny most surely will be squarely *on* the agenda of future, frankly eugenic medical interventions—much of it done while progeny are still embryonic. Blish's world hovers eerily within Ben's.

---

[8] Despite the apparent promise of such new potential therapies as individually designed treatments utilizing a patient's own immune system.

What is novel about molecular biology and genetics is that very little, perhaps nothing will again be regarded as automatically beyond the social or medical pale. Everything, in short, formerly beyond the limit may soon be up for review, new study-designs, and possible if not yet probable reversal, correction, even replacement if need be. At the heart of this vision, it seems, we ought not fear being on a slippery slope, but should instead welcome, even relish the novel vistas, prospects and exhilarating ride—which are potent indeed.

(b) In Mawer's novel, Dinah is deeply ambivalent toward Ben, at once attracted and grateful, yet repelled—not unlike many others of us when we are in the company of the likes of Ben—who, on the other hand, is not only extremely nice to Dinah but goes out of his way to help her get through the genetics class. Why then are the Bens of the world so disturbing? Dinah is beside herself when she passes the course, spontaneously kisses him, then promptly tries to take it back. Befuddled, yet on fire and riveted by "*she* kissed *me*!," Ben tells her he loves her, and her response? "I knew you'd do this...can't you see it's impossible?" to which Mawer has Ben reply: "Of course it's impossible. It's the impossible that attracts me. When you're like I am, who gives a toss about the possible?" (Mawer, p. 52) He then says what she cannot bring herself to say, that he is "hideous and deformed, a freak of nature, and people would stare." As if the rest of us were not chance creatures of accidental genetic fusions as Ben or other dwarfs!

There is at just this point something still left unsaid, unspoken, perhaps even unthinkable even as Ben himself tries to think and say it—or, perhaps, it can be spoken only because the one who is unspeakable, Ben the dwarf, says it for her. Why, we must wonder, is it so hard for her to say what she actually thinks, and to say it directly to Ben? Isn't utter honesty called for? Why would it be difficult for any of us to say it to someone like Ben? Why do we hesitate, when on the other hand what is unsaid is if anything utterly decisive for what we then think about and how we act toward Ben the dwarf?

(c) When discussing using human subjects for research in 1865, the famous scientist, Claude Bernard, did not mention nor did he presumably intend to mention anything like informed consent. He wrote, rather:

> It is our duty and our right to perform an experiment on man whenever it can save his life, cure him, or gain him some personal benefit. The principle of medical and surgical morality, therefore, consists in never performing on man an experiment which might be harmful to him to any extent, even though the results might be highly advantageous to science, i.e. to the health of others. (in Katz 1992, p. 229)

Commenting on this passage, Jay Katz notes "that Bernard spoke about 'our duty' and 'our right'; he said nothing about research subjects' consent," much less their 'duty' or 'rights'. And, continuing to reflect on this remarkable passage, Katz seems taken aback by his realization that

> One question has not been thoroughly analyzed to this day: When may investigators, actively or by acquiescence, expose human beings to harm in order to seek benefits for them, for others, or for society as a whole? If one peruses the literature with this question in mind, one soon learns that no searching general justifications for involving any human beings as subjects for research have ever been formulated. ... Instead, in the past and even

now, it has been assumed without question that the general necessity for experimenting with human beings, while requiring regulation, is so obvious that it need not be justified. . ... I do not contend that it cannot be justified. I only wish to point to the pervasive silence. . .and, more specifically, to the lack of separate justifications for novel interventions employed for the benefit of future patients and science, in contrast to those employed for patients' direct benefits. (Katz 1992, p. 231)

At the heart of this, Katz says, is "a slippery slope of engineering consent," one that leads "inexorably to Tuskegee, the Jewish Chronic Disease Hospital in Brooklyn, LSD experiments in Manhattan, DES experiments in Chicago"—and many, many others might be added—all of which are "done in the belief that physician-scientists can be trusted to safeguard the physical integrity of their subjects" (Katz 1992, p. 231). As the former editor of the *New England Journal of Medicine*, Franz Ingelfinger, once insisted: "The subject's only real protection, the public as well as the medical profession must recognize, depends on the conscience and compassion of the investigator and his peers" (Ingelfinger 1972).

It might be said, of course, that the physician-scientists involved in these events, like those who conducted the experiments in the Nazi concentration camps, the Gulag Archipelago, Willowbrook, or others were perverse or even evil persons. Science and medicine are value-neutral; they are "intrinsically benign," it might be said; (Weissmann 1982) evil actions stem not from science but from individuals who are evil, or who do evil things, because of the ways they use science and medicine. To suggest otherwise, Katz verges on saying, would be to court something scandalous—if not unspeakable or unthinkable, then surely repugnant, and that would be something awful, appalling even, quite as much as engaging in an act of incest.

Three things are clear. (a) In the new genetics, very little seems beyond the limits of newly possible interventions designed to correct, re-figure, conquer, or replace—most of all before flawed genes can do their inevitable work. (b) Reflecting on Mawer's narrative about Ben Lambert, something unspeakable emerges as, somehow, connected to the first point: that we dare not say what we truly believe about individuals such as dwarfs—until and unless, that is, something can be done to correct or ameliorate phenomena such as achondroplasia—and, it should be added, only so long as we remain oblivious to our own accidental conditions. That silence is surely just as puzzling as (c), that "pervasive silence" which puzzles Katz.

There seems nowadays to be the possibility, at least, of a sort of license for genetic medicine to try and undo, replace, or even transcend nature and natural evolution (as Eccles in fact proposed), to re-make Ben, because being Ben is profoundly offensive—in much the way the 'feebleminded' were regarded by Darwin:

With savages, the weak in body or mind are soon eliminated; and those that survive commonly exhibit a vigorous state of health. We civilized men, on the other hand, do our utmost to check the process of elimination; we build asylums for the imbecile, the maimed, and the sick; we institute poor-laws; and our medical men exert their utmost skill to save the life of every one to the last moment. (Darwin 1874/1974, pp. 130–31)

Saving the imbecile, the severely disabled, the simpleminded, the hopelessly confused and non-productive—even encouraging them to reproduce—can only be "highly injurious to the race of man," Darwin believed, and eventually leads to the degeneration of "man himself." The sensible thing for Nature, or God, or whatever set evolution in motion in the first place, would have been to prevent such individuals from reproducing. Since that did not happen on its own, so to speak, Darwin and his legacy took it on themselves to do it by inspiring, if not recommending, various sterilization laws to prevent the feebleminded from reproducing. It then naturally follows, Kurt Bayertz argues, that with this striking failure of "natural selection," the grounds were well-prepared for the more recent proposals for deliberately controlled experiments to produce more useful citizens, (Bayertz 1994, pp. 42, 44) precisely as Eccles, Burnett, and other geneticists and molecular biologists had proposed, using whatever means necessary. Eugenics follows closely.

Still, we must wonder about that "pervasive silence" by the research community. As Katz sees it, it leads to "a slippery slope of engineering consent" for research projects, and this in turn "inexorably" leads to the dreadful perversions of Tuskegee, the radiation experiments in the United States with whose outrageous aftermaths we are still living but which took so long even to acknowledge publicly. How could any of this happen? How can any of it be understood? How could any physician in the restorative, Hippocratic tradition ever be caught up in such deliberate designs that not only ignore, abandon, and literally overlook individual human beings, but even more to do so in the name of science and medicine?

9.6 Speaking to the Unspeakable

Rheinberger wrote that, taking off from the incest taboo: "Just as the incest prohibition became the scandal of anthropology, so has the commandment of truth become the scandal of the sciences of natural things," (p. 259) including the human body. What could he mean by this? Is it that, say, with the "deliberate 're-writing' of life" (p. 253) that he takes as the basic aim of the new genetics, there is then introduced what on the other hand is capable of fundamentally altering the very life that conceived then carried out the new genetics?—such that, perhaps, the very possibility and ability of future generations to do this as well can and perhaps will then been made impossible? Because we *can*, are we then free to try and cancel the same sort of freedom of action of future generations? Does that *can* imply *ought*?

Or is it like the Pasteurian program a century ago, also cited by Rheinberger—a program that rejected the entire question of theories or goals but thought of means merely—which is precisely what, having established the Genome Project, is now being embraced by the molecular biologists and project managers of the National Institutes of Health and the Department of Energy? Rheinberger quotes Latour to make his point: those Pasteurians, not themselves especially potent in political terms, nonetheless

> ...followed the demand that [their own weak] forces were making, but imposed on them a way of formulating that demand to which only [they] possessed the answer, since it required [men and women] of the laboratory to understand its terms. (p. 254) (Latour 1988, p. 71)

Nothing, Rheinberger insists, "could describe the political moves of James Watson, Walter Gilbert and their combatants better than this quotation." (p. 254) Is this then the scandal: the spectacle of this remarkable finesse of politicians by, of all people, scientists, who are typically thought to be politically ineffective but who yet secured immense funding for a scientific project riding on the back, it seems, of what Rheinberger gently calls a "misunderstanding?" He means, I gather, that genetics is not in the first place so much about diagnosis or even cure of disease, as it is about improving people (or some of them) by controlling and "exploiting" (Eccles' term) human and animal evolution—aims which, not sitting well with a largely uninformed public, must then be somewhat hidden behind stated aims that are valued by that same public, such as treatment for awful diseases like the cancers.

The new genetics is for all practical purposes capable right now of serious control of human reproduction; experiments with animals since the early 1980s demonstrates that the same can be done with humans. Is the scandal then—what either should not, cannot, or will not, be openly admitted—that only those with the power to control the knowledge and exploit the technology will do so, and they will never let the truth of what's happened be known—a type of potent, silent priesthood, echoing Orwell's *1984*? Is the point that scientists should or must give politicians the ammunition needed for their re-election—march genetics out under the banner of changing medical practice by finding ways to cure disease— and the money-mill will open wide? And that the rest of us will not be able to know before or after the fact what actually goes on behind closed doors?

9.7 Politics, Power and the Loss of Norms

This may be at least in part what Rheinberger is saying with his talk about scandals. But for his case to be well argued, it seems to me, there is something else that needs an accounting. How and why is it that such wide-spread suspicion, distrust spreading to everyone and everything has come about, (Pellegrino 1991) especially toward one of the last bastions of social prestige and authority, medicine (and its underpinning bio-medical sciences)? And why is there distrust, even cynicism, concerning those people who can and will actually control procreation?[9] Is this not just as clear regarding the theme here and it has become obvious on such topics

---

[9] The distrust is open in, for instance, Leon Kass' rejection of human cloning (Kass 1997).

as global warming? Is this distrust not grounded in the same ignorance of science and scientific method?

Barbara Ehrenreich caustically noted in her editorial for *Time* on the occasion of the first (although mistaken) announcement in 1993 that human cloning had been achieved: we should, she wrote, be very apprehensive, not about twenty first century technology—which promises the kind of genetic cloning Blish had forecast in his cunning novel, technology with which to "seed" the stars—so much as putting such potent technologies into the hands of twentieth century capitalists, whose money, after all, pays for such adventures. If not about the scientists, then, whose research results in the feared technologies, then distrust of the genetic engineers who will put the theories to work; and if not them, then toward those who provide the funding for the enterprise, or possibly those with positions in policy formulation and enactment.

Is this passion for control, for epistemic and political power, then, the real scandal? If so, then Rheinberger's point about science and the loss of truth makes a good deal of sense— the crucial point isn't what you know, but who owns the means and products of research, not unlike Marx's reflections on capitalism in the nineteenth century, indeed the logical extension of his thought. Which, to be sure, may well be a scandal in the sense that, if present trends continue, the very sciences which proudly parade a commitment to truth would, in their constant and upward-spiraling escalation of costs (and search for escalating financial support) (Rescher 1982) be for sale to the highest bidder and thus undo that very commitment to truth. Is the scandal, then, that once on this fateful path, its course is inexorably set, like the very best of slippery slopes, even if concealed by nice, kind words?

I think at this point of the inevitable reminder: the astounding Grand Inquisitor scene in Dostoyevsky's *The Brothers Karamasov*, where, after hearing Ivan's tale of the return of Christ and the priest's objections to that, Alyosha cries out with his riveting question: is anything then permitted? Is nothing forbidden? Can anyone then do anything they want, simply because at this or that moment they by chance happen to want it—and can pay for it?[10]

It would appear that underlying modern medicine's impending realization of its ancient dream to improve the human condition is set deeply within something that resists being expressly spoken. Which may be the actual scandal, for must we not wonder about the wisdom of the choices that will, it seems, inevitably be made by those who will make them simply because they alone understand the technologies, or have paid for them? We must wonder, too, with Hans Jonas, about efforts to rectify and alleviate the "necessities and miseries of humanity" by "inventions" in the manner of Bacon (i.e. through technology), at the same time so conceiving knowledge that no room is left for what can alone provide guidance, a knowledge of "beneficence and charity" (Jonas 1966, p. 189).

---

[10] Itself a stark reminder of what Edmund Husserl pointed out at the very beginning of the twentieth century in his 1910 essay in the journal, *Logos*: "Philosophy as Rigorous Science." (Husserl 1965).

No matter how well Bacon actually understood the necessity for moral guidance for that "race of inventions," his project succeeds only in creating a powerful paradox, since neither theory nor practice in this usage contains or can say anything about such goals or moral governance. For, neither beneficence nor charity "is itself among the fruits of theory in the modern sense," nor is "modern theory...self-sufficiently the source of the human quality that makes it beneficial." Indeed, Jonas argues,

That its results are detachable from it and handed over for use to those who had no part in the theoretical process is only one aspect of the matter. The scientist himself is by his science no more qualified than others to discern, nor even is he more disposed to care for, the good of mankind. Benevolence must be called in from the outside to supplement the knowledge acquired through theory: it does not flow from theory itself. (Jonas 1966, pp. 194–95)

Emphasizing that the prospect of genetic control "raises ethical questions of a wholly new kind" for which we are most ill-prepared, Jonas later urgently suggested, "Since no less than the very nature and image of man are at issue, prudence becomes itself our first ethical duty, and hypothetical reasoning our first responsibility" (Jonas 1984, p. 141).

H. T. Engelhardt, Jr. came to much the same conclusion about modern medicine. Echoing Jonas, he wrote: "Man has become more technically adept than he is wise, and must now look for the wisdom to use that knowledge he possesses" (Engelhardt 1973, pp. 451–52). Recall T. S. Eliot's incisive, thundering questions: where is the knowledge we have lost in information? and where the wisdom we have lost in knowledge? Jonas went on to emphasize that we are therefore

constantly confronted with issues whose positive choice requires supreme wisdom—an impossible situation for man in general, because he does not possess that wisdom, and in particular for contemporary man, who denies the very existence of its object: viz., objective value and truth. We need wisdom most when we believe in it least. (Jonas 1974, p. 18)

It is not so much that we are continually threatened by one or another slippery slope. Rather, I believe, being on a slippery slope is precisely the human lot, what it means to be human, at least since Darwin and in particular since the disasters of the twentieth century—the Dreadful, R. D. Laing asserted, has already come about, (Laing 1967) and I think it bears all the signs of Dostoyevsky's breathtaking "anything is permitted," and nothing forbidden.

9.8 Thinking About Birth, and Beyond

Even at this point, I have a sense that there is something else still lurking in the darker corners. As mentioned, at issue in the Genome Project is a fundamental philosophical-anthropological issue: not only how self is at all known and experienced, but whether there is self *at all*, much less 'person,' or instead, merely genetic information encoded in or on strands of DNA/RNA nestled within any individual's body cells. Walter Gilbert's excited pronouncement, "Here is a human being; it's me'!," etched on a CD, is a challenge none of us can ignore. Does it not pose very much the same question of scandal that Rheinberger dares us to face?

To help make my way through these complex matters, I think it is helpful to dwell for a bit on several peculiar and quite technical passages in the work of Alfred Schutz. One appears in his critical review of Husserl's understanding of intersubjectivity; the other in his intriguing article on Max Scheler.

(a) After insisting in the first that intersubjectivity is a "given" and not a "problem" to be solved, Schutz insisted, "As long as man is born of woman, intersubjectivity and the we-relationship will be the foundation for all other categories of human existence." Accordingly, he continued, everything in human life is "founded on the primal experience of the we-relationship," (Schutz 1966, p. 82) which, though he didn't explicitly say so, must surely be the experience of being "born of woman." Since *all* "other categories of human existence" are founded on this primal experience, our being with and among other people was for Schutz "the fundamental ontological category of human existence in the world and therefore of all philosophical anthropology" (Schutz 1966, p. 82).

In the Scheler essay, Schutz's words are equally fascinating. He first pointed out that there is one taken for granted assumption which no one for a moment doubts, not even the most ornery skeptic: "we are simply born into a world of Others," for even that skeptic, the one who doubts the existence of other people, was in the first place himself born and raised by some of those very other people! Then he said: "As long as human beings are not concocted like homunculi in retorts but are born and brought up by mothers, the sphere of the 'We' will be naively presupposed" (Schutz 1967, p. 168). Here, too, it is reasonable to surmise that what is "naively presupposed" is precisely that "primal experience" of being borne by a woman (I need to add), and "born of woman" and (he added here) being raised by mothers as opposed to being "concocted...in retorts."

What I want to pick up on is the idea that "being born of woman" constitutes "the" (not merely "a") "fundamental" ontological and anthropological category of human life. It is curious to note first that few philosophers have thought it necessary or, I suppose, fruitful to focus on this phenomenon of "having been born of woman." Reflections on death and dying are plentiful; those on birth, being borne and then born or 'worlded,' are oddly lacking. Still, if we consider this—even if, as Schutz also said, we can get at it only indirectly, through other people (Schutz and Luckmann I: 1973, p. 46)—still, my having been borne and born are surely as constitutive of my life as is my going to die.

Schutz did not probe this phenomenon much beyond these scant references. Still, his words have to be taken quite seriously, for in a clear and compelling way it is the primal experience of being (or having been) borne and born that constitutes the crucial other side (other than death) of the central experience of growing old together, and of our being-with-one-another—of what he terms the "tuning-in relationship" or of intersubjectivity. We could not experience ourselves as growing older together, if we did not begin to be—so to speak, if we did not come at some always-already-ongoing time in our lives to find ourselves as having-already-been-thrust-into life: birthed and thereby 'worlded.'

To be born as human, but more specifically, as myself, is *to have received* life, *to have been given* my life—the first and fundamental sense of *gift*. And in this, it seems clear as well, lies the fundamental *paradox of freedom*: while a prime condition of morality (choice, responsibility, etc.), I do not choose to be free but, as Sartre saw, I am not free to choose to cease being free. Hence, an ethics that focuses on giving is seriously incomplete without a complementary reflection on an *ethics of receiving*; the latter may indeed be the more important phenomenon, if only in view of its having been so oddly avoided by so many.

The primal Other, in short, is the *mother*, the one with whom each of us in the first instance grows older, in Schutz's words; and the initial and primal *place* or habitat is her literal body, her womb. She is *the one who gifts* me with myself and is progressively *the one who gifts* me with herself. From her I receive my here-and-now presence but also my culture, history, world, mainly through giving the key stories by which I come to know myself.

I am not only, then, a "being-toward-death," (*Sein-zum-Tode*, Heidegger) but surely just as fundamentally a "being-*from*-birth"—indeed, in a sense my being is always-already a "being-*before*-birth," being already within the mother's body; this is thus the originating sense of my becoming. What and who I *am*, is what and who I in multiple ways *become*, and this is first set in motion in the essentially mysterious and accidental ways of every birth.

This returns me to Ben.

9.9   One More Indirection

Mawer has Ben reflect when he's in the passionate moment of wondering how it was that he ever came to be just this specific person, this Ben the dwarf. There simply is no way to know the why or how just one specific sperm made its way into one specific ovum, nor the countless accidental splittings, changings, connectings, shiftings, turnabouts, of both Ben and his mother as she bore him from the tiniest of the tiny into birth, and beyond, into himself. Even were there to have been an *in vitro* infusion of a pre-selected sperm—'get that one there, Shirley. . .no, not that, *that* one over there. . .'—how account for, how make understandable, what consti-tutes just *that* life, *that* unique life which then, if all goes well, becomes just that unique individual, Ben Lambert, dwarf? Can that be somehow said? Stated? Explained?

I hope another indirection will be permitted, as I try to make these matters even clearer. In one of the narrative I've written, taking off from actual clinical life,[11] one

---

[11] "The Indomitable Rachel Bittman," a story, in *Langdon Review of the Arts in Texas* 10 (2013–2014), pp. 68–79, preceded by "Why I Write".

of the characters is an elderly man whose wife is in the final stages of Parkinson's Diseases. He recalls her talking to him several months before she began to become more stone-like than flesh.

He recalled what Rachel said to him after he started once again to talk about trying to get her into some sort of clinical trial, but couldn't. He recalled how "her voice picked up a notch, then she said, 'Cy, you've got to stop. You know, well as I, it's just too late for me, and even if it weren't too late, and it is, it's still only '*might* be and *might* help,' never '*will*' not even '*can*' help. That's God's own truth, Cy, God's own way of doing things.'"

"'God's way'?" I wondered aloud. "What'd she mean?"

"Well, Rachel believes mightily, I've got to say, that God is good and true with us humans, with every creature. And 'God,' she said, 'well God has His reasons for things, or at least He has his understanding, which you and me, we just have to set our minds and try to figure out. I think it must be something like this,' she said, 'God created *everything*, without God none of us, none of this would be, at all.' And I remember her sweeping her arm around; she could still do it then, a little. 'But God created us with free will, Cy. Free will, you understand? We're always free to choose, free to recognize or ignore His works and, fact is, Cy, we *have* to choose. There's just no other choice since God designed us as creatures that choose, who freely decide and choose; I mean, you see, even if we try to choose not to choose, we're still choosing, still using our free will. At first we were all in His hand, and then, when we're born—well, He just let us go out of His hand, and then it's all up to each of us to find our way, choose our path, and then just get on with it. But He never gave us any guarantees.'"

"She was really committed to this?" I wondered.

"Yes, least, that's about what she said. What she meant was that I had to stop getting my wishes ahead of the facts. 'Cy, you just let your hopes get ahead of what's possible; if only wishes were horses, you'd really be on a ride!' She was just so smart that way."

I nodded so he'd continue.

"I remember telling her that there always a chance something will work. And she just jumped back at me: '*Chance*! Exactly, Cy, just a chance! That's what I think God is all about,' she said. 'When He let us go out of His hand, chance took over; it's the way He does things, once he made us free to choose for ourselves. Well, it's like we just can't live with uncertainty and chance, and just want to control, be sure about everything!'"

"Wow," I muttered, "that's impressive."

"Well, you know, her words were powerful. I never thought much about that before Rachel came back here, but then I started getting as much stuff to read and study as I could, anywhere I could."

"Yes, I think I know how you feel about such things."

"Well, think about it, right now: how did she come down with Parkinson's? We just don't know, though there's lots of guessing about it, about the genes, or about things in the environment, poisons and such put there by dumping or whatever. But, like Rachel says, her getting Parkinson's was 'just one of those things that happen, no more, no less. Can't blame God or anyone or anything.'"

9.10 Of the Scandal, Chance, and God

Now, these reflections suggest not so much that the new genetics places this entire, awesome reproductive process at risk, as Rheinberger and many others seem to suggest; nor do the novel genetic techniques and theories threaten my 'who I am' and the variety of foundational relationships among us (father, mother, son, daughter, etc.), as Leon Kass insists (Kass 1997). Rather, it is *my being at all* that is at issue, for just this is now placed in a radically new light, and in this there may be a true scandal: *that* I am *at all*, *that* I have come or been brought into being (into life) neither through my own action or choice, nor through anyone's decision, while yet being born free to choose from that point on. Being a creature with 'free choice,' I am yet made to be, so to speak, without the least choice on my part—any more than Ben chose to be a dwarf, or Rachel to come down with Parkinson's.

Nor did Ben's parents choose Ben, this unique individual. Perhaps they had wanted *a* baby, but *his* coming on the scene, the unique *Ben*, is wholly outside any parents' or anyone's ken, foreknowledge, or choice. Being a baby—being just *this* baby—is always and essentially a surprise—to itself and to its parents. But the reverse is also true, for Ben no more chose his parents than they chose him. Hence, for Ben to be what he is, to be himself, is to be an ontological surprise. *He* is an accident (the 'accident of birth') that embodies chance in its purest form, though being himself is not only that.

What is scandalous about that? At one point in Mawer's deeply ironic novel, Ben succeeds in sequencing the genes which, incorporating a single, apparently trivial error in a single base pair in "this enigmatic, molecular world," (p. 197) likely eventuated in him, Ben Lambert. That so-called "genetic error" involves a "simple transition at nucleotide 1138 of the *FGFR3* gene" (p. 198) that, in the dark recesses of his mother's womb and impregnated by a single sperm, mutated into what eventually became Ben. A single mistake in the 3.3×10^9 base pairs in his genome, one mistake, one substitution of guanine for adenine, in the trans-membrane domain of the protein—that part which fits through the cell membrane—and the result was Ben. Is this not a scandal: the sheer, accidental fact that, of all the millions of pairings along those snaky, helical arms and spiraled columns of deoxyribonucleic acid busily replicating, churning out proteins, those building blocks of life; a single exchange, a single letter error, and there's Ben, the achondroplastic dwarf, that gnarled, disfigured "monster" who despite everything is a genius and, more, loves Jean? And Jean, too, accidental outcome of the same sort of sinewy organic and sub-organic workings, tries mightily to love him, too, but in the end has to confess that she just can't—she can neither leave her husband, nor be with Ben. Neither would be fair, as she later says (p. 179).

And here I must pick up on Herbert Spiegelberg's insight: (Spiegelberg 1961, 1974) we must note that, having no choice in his birth—not even *that* he will be born—yet Ben and each of us as we grow older assumes the prime responsibility for ourselves, for what each of us then becomes; choice has entered and begins to decide and design, whether cleverly or not. Save for that initiating happenstance, each of us is responsible for whatever may eventuate. At some also unchosen point Ben gradually emerges from a globally undifferentiated entity at birth we name and celebrate as 'baby.' From the same playing out of chance, Ben could just as easily *not* have been born, hence not *be* at all, —or if born then born without that chance mutation, and for any number of incalculable reasons themselves as accidental as that the multiple biological processes and timings managed to eventuate in his birth. But from then on, it is *his* life, whatever he may subsequently do or not do about that: he, Ben, is the continuous outcome of *chance* and *choice*. Even more, beyond all that, being born as 'me' with its unchosen accoutrements is, Spiegelberg is anxious for us to understand, the purest kind of "moral chance" and is therefore utterly undeserved: there is no "moral entitlement" to what I happen to be, whatever the station of my birth, no more than what I biologically inherit is something to which I am entitled. All of which is just what Rachel Bittman tried to get her husband, Cy, to accept, understand about her journey with Parkinson's.

The phenomenon of moral chance seems quite essential to having been born of woman, mother—nor, I strongly suspect, can there be any ontological or theological accounting for that uniqueness which each of us is already before and at birth. As I think about Ben, it seems to me outrageous that he (and each of us) was, choice-lessly, saddled with being *him*; it seems altogether scandalous, moreover, that he (and each of us) should have either 'advantages' or 'disadvantages' simply because of the numerous accidents that eventuate in birth. But precisely the same is true for each of us, both in our biological wherewithal and in our initial stations in life (which family, which place, which time). Is it not outrageous that any of us is born at all, with all of what we are and who we become?

All that is a kind of prologue to something equally if not more puzzling still. This arises from the choice Ben faces when Jean, who has already returned to her infertile husband, asks Ben to use his sperm for the *in vitro* fertilization she has already asked him to perform. He agrees. Later, he checks the fertilized eggs, has an associate gently suck up each embryo in turn, while he himself does the PCR amplification. He determines that embryos 3, 5, 6 and 7 are unaffected; they show no 'misspelling' of adenine by guanine. But he also determines that 1, 2, 4, and 8 show that very mutation; adenine has been replaced with guanine, and achondroplasia is irreversibly on the way.

By chance, four 'normals' and four mutations have come about as dwarfs-to-be—if allowed to be all. What should Ben do? Note well: he *can* actually choose one or more embryos to implant; he can select which, by implanting, will be allowed to grow into a baby and, it may also be, become another genius like Ben. But also note that none of these embryos have any choice in the matter.

Is this then the situation God confronts when He goes about the business of human birth? Should Ben 'play God?' Mawer sets the scene: "Benedict Lambert is sitting in his laboratory" with eight embryos in eight little tubes. "Four of the embryos," he reflects, "are proto-Benedicts, proto-dwarfs; the other four are, for want of a better word, normal." How should he choose? And is his choice, whatever it may be, acting or 'playing' like God? The narrative then continues:

> Of course, we all know that God has opted for the easy way out. He has decided on chance as the way to select one combination of genes from another. If you want to shun euphemisms, then God allows pure luck to decide whether a mutant child or a normal child shall be born. But Benedict Lambert has the possibility of beating God's proxy and overturning the tables of chance. He can choose. Wasn't choice what betrayed Adam and Eve? (p. 238)

So, Ben would not be 'playing God' in the least; if he were to do that, he would have to find a way to let chance work its way. But Ben can choose, and when he makes the choice, does the deed, and the baby is on its way, Jean telephones to ask him what he did: "Is it all right?" Which embryo was implanted? The conversation heats up as Ben evades and dodges, knowing full well what he has already done, and cannot now undo, and doesn't want to tell her. But Jean pleads with him, to the point where he grows angry "at the docile stupidity of her, at the pleading, whining kindness of her, at her naïveté. 'Well, you'll have to wait and see, won't you?' I said to her." Then he hangs up. Was that in any sense fair? Was it just?

In the narrative about Rachel Bittman, she is at the other end of life, desperately seeks someone to "help" her. Help her do what? Well, she wants someone to help her die; at the time this is really clear, she has little left of her life, she realizes. When she's left with nothing else, what then? Still alert to what's happening to her, to what will happen before long, she wants to be helped to die, whether by use of opioids or barbiturates, some way to ease the pain and speed her dying. But wouldn't that be pretty much what Ben faces, at the other end of life? If God could not do what Ben faces, could He do what Rachel wants? So what is it that gets Ben's anguished attention in the laboratory? What would the one who 'helps' her die be doing?

Spiegelberg, as I understand him, in a certain sense addresses just this sort of issue, namely, that there is a much "deeper sense of justice" and "injustice" than is usually discussed, something he says that is genuinely "cosmic," at the core of our lives. His point is that since, (1) *"undeserved discrimination calls for redress"* and since (2) *"all inequalities of birth constitute undeserved discriminations"*, he concludes that *"all inequalities of birth call for redress,"* therefore that *"inequality is a fundamental ethical demand"* (his emphasis) (Spiegelberg 1961). This demand, moreover, is intrinsic to the phenomenon. If that is so, on whom does the responsibility for redress fall? But is this true for Ben? Will it eventually be true of his the child that will eventually be born from Jean's body? Is it true for Jean, too?

Does Ben's "inequality of birth" call for redress? Indeed, is it not rather the case that, while some of us may well feel how profoundly unjust it is that Ben was born, we cannot avoid the awesome question: was Ben's birth *unjust*? Even if it were, does that imply a demand for redress? If so, who redresses, and what, exactly, can be redressed? And, finally, what exactly is "unjust," cosmically or otherwise? Is it that, through no fault of his, Ben is a dwarf? At the same time, however, each of us must know that who and what we are did not come about through our own choice either—and just because of that each of us, dwarf or 'normal,' is essentially in the same quandary as Ben might be. The same issues are obvious as well for Rachel's quandary.

But is it the same for the embryo Ben must chose to implant? Or for the one who decides to help Rachel on her way into death?

9.11   Beneath the Scandal that I Am Myself

Each of us is born with some initiating conditions that is utterly unchosen, undeserved and, surely, an inequality of the first order. Does Spiegelberg's passionate focus work here? I think not, and *that it does not seems outrageous, a real scandal*. What happens after the brute accident of birth, that's something else, something with respect to which this or that course of life may or may not ensue, with responsibility properly meted out for these as for all other people. But is it the same for the bald, brutal fact of initial biological, familial, and in general existential wherewithal? It does not make sense to talk about "unjust" and "redress" here, and *that it does not make sense* is scandalous; each of us is aware at some point and in some way of what Spiegelberg terms 'unjust,' but there is nothing any of us can possibly do about that!

On the other hand, it strikes me as clearly wrong to allege, for oneself (if born as Ben) or for another (Rachel's 'helper'), that 'God did it' and is responsible, hence must be called to account for the offense! Ben reflects, after donating his sperm:

> [For] what is natural? Nature is what nature does. Am I natural? Is superovulation followed by transvaginal ultrasound-guided oocyte retrieval natural? Is *in vitro* fertilization and the growth of multiple embryos in culture, is all that natural? Two months later. . .I watched shivering spermatozoa clustering around eggs, *my* spermatozoa clustering around *her* eggs. Consummation beneath the microscope. Is that natural? (pp. 214–5)

Precisely here it seems is a true scandal: each of us is born and in the *fact* of being here at all, much less in the way and how we each are, we are initially what and who we are thanks to a plain throw of the dice, the sheerest of chance, locked in and by the accident of birth. As Schutz apparently appreciated, each of us is borne and then born, "not concocted like homunculi in retorts but born and brought up by mothers," and in this we each are without exception accidents, here on this earth. And this, I think, openly reveals the brazen hubris of Gilbert, of Watson, of Eccles and their promises of control in a world governed to the contrary by the genius of chance.

9.12   Am I Me Solely Within You? Are You Solely Within Me?

A way to appreciate what's so compelling about Schutz's otherwise only isolated suggestions is to consider them in light of that theme *du jour*, human cloning.

A cloned human infant, of course, is not the same, in Schutz's terms, as a being "concocted in retorts," although when he used this then common term, it probably amounts to much the same. In any event, it is clear that a cloned human being hardly ceases to be human simply because it is cloned. As even the most hard-nosed genetic determinist knows perfectly well, moreover, in the case of cloned individuals all that's different is that they share most (one cannot overlook the fact of mitochondrial DNR/rDNA, nor the results of continuous chance mutations) of the same genome, by design and deliberate plan rather than the usual delightful way—as in the case, absent the deliberate planning, of naturally occurring identical twins (who also, of course, share the same genome).

Wilmut's method, (Wilmut et al. 1998; Report 1997) as is well-known, involves the nuclear transfer of genetic material from an adult cell to an egg taken from another adult, which is then implanted into a third adult's uterus. Here, it is clear, there is not only deliberation and planning, but the reproduction itself is asexual, which is the very thing that worries Kass and others. To be sure, a non-human uterus (which Schutz somewhat naively calls a "retort") might conceivably be developed. It has already been demonstrated that human genes can be spliced into the cells of certain animals (though there are, of course, potentially serious problems with this[12]), to produce certain human proteins. Eventually, human tissues and even solid organs might well be produced. Could a full human fetus be similarly developed? It is not as yet clear how or whether that question could be answered.

One thing is in any case very clear. The human fetus within a human uterus exists and has its being solely within a continuously developing context or network of intimate interactions with the mother and even with other individuals, although much of this is still but poorly understood. In any case, it is thanks to that developing network that what we otherwise term 'fetal development' is truly 'human development' in its earliest and clearest form. I mean: to be human is to *become* human; and becoming human requires a sequential development whose primary characteristic is that each of its stages is or involves a continuous complex context of interrelationships with a highly specific Other, the mother.[13] Each of us is at the outset of our lives truly *always-already-with* mother; we are *always-already-within* the literal embrace of her body, from the earliest stirrings of semen-penetrated ovum to the full infant immediately prior to birth.

Schutz understood with remarkable if also undeveloped insight that the prime phenomenon here is *receiving* life, being gifted with myself by the mother. What he

---

[12] In particular, the potentially lethal consequences from alien viruses and bacteria.

[13] I have used a neologism to capture this complexity: complexure.

did not probe were the implications of the "primal experience"—and it is just this phenomenon that comes into question again with the advent of human cloning. He also seemed to have understood that without that ongoing biological process of pregnancy, it would be profoundly questionable whether any 'outcome' could conceivably be 'human.' If we suppose it were possible for there to be some sort of artificial womb and placenta housed in some laboratory somewhere—a completely novel sort of 'intensive care unit' from the earliest moments of impregnation on—and suppose further that an appropriately cloned or semen-penetrated egg could be implanted in it, we would then have to contend with the really difficult question implicit in Schutz's words. Would an "homunculus in a retort" be 'human' if it did not issue from impregnation, implantation, and fertilization, and was not allowed to stay and grow in mutual relationships within that primal human environment, a female human being, its mother? If what I have suggested is correct, nothing but a "homunculus" could possibly emerge from such a retort. To be human, to repeat, is at the very least to become human, and becoming human in stages along life's way requires that temporal sequential development within and nourished by another human body.

Thus, when Jean Bethke Elshtain asserts, in what she says is her own "nightmare scenario" (cloning human beings to serve as spare parts for, one presumes, other adult human beings), that "cloned entities are not fully human", (Elshtain 1998, p. 182) she is quite evidently mistaken, her "nightmare" nonsensical—*unless* such an entity were conceived and carried in at least its initial journey outside the mother's womb. The uterine environment, in other words, strikes me absolutely essential, though it is not all that is essential, for such an entity to become human (Zaner 2003).

The risk of cloning, then, is not some supposed threat that it will erase the unique individual or its network of relationships with others (mother, father, son, etc.). It is, rather, the loss of that for and in each of us, which comes to be within and by means of my relating to you and you relating to me: it is, ultimately, *we*, you and I, who are at risk. This is not true of natural identical twins, for they are both nurtured and enabled to grow toward birth within the mother's body, and in that intimacy come to be as and who they are—clones both of them and none the worse for that. When born, however much alike, they are yet destined each to be that self each is solely in relation both to mother and to others, especially to the twin—who are each also self in relation both to one another and to the twin.

9.13   Concluding Reflection

The fact of the accident of birth gives a quite different sense than usual to the idea of the 'slippery slope' that has had such attraction over the past four decades. The horror at the bottom of the slope, it must now be clear, is that there simply is no bottom, nothing solid whatsoever, only a steady, slippery slope initiated before the accident of our individual births. It is, in a word, our human condition—to be always in search of firm or firmer footing than that presently at hand, and perhaps inevitably to be disappointed in our failure to find it. In this respect, that fabulous slope is not unlike what Albert Camus brilliantly stated in his great work, *The Myth of Sisyphus*. His words allow me to bring this long refection to some kind of conclusion.

Camus had the great courage to say, out loud for all to hear and see to read, that any appeal whatsoever to transcendence and absolutes (that supposedly firmer footing that moves so many of us at times of radical uncertainty) can only be "absurd." Such an appeal is but one of the machinations by which control is sought; it is but a way to try and ensure that the one who asserts the transcendent or the absolute also asserts that he or she knows better than anyone else what's good for all the others. As if there were something absolute; as if, even if there were, such an absolute would be the truth of who and what we are; as if, even were that coherent, this or that finite human being could apprehend it surely and doubtlessly; and as if, apprehending it in one grand sweep of thought innocent of every infelicity of being a specific, error-prone, historically bound individual, this were not the height of hubris (Winner 2004).

Camus' point, or some key part of it, is that such schemes are beyond our capabilities. Such appeals to some sort of higher ledge of authority—available to no one else and from which to pronounce judgments on the rest of us—are but tacit signs of dread and doom, of the deep uncertainty and chance that constitute our condition as human. "I want to know whether, accepting a life *without appeal*, one can also agree to *work* and create *without appeal* and what is the way leading to these liberties." And this, set out as starkly as the sun-blistered sands in that striking, colorless beach in *The Stranger*, may be the sole way genuinely to reclaim our lives. "I want to liberate my universe of its phantoms and to people it solely with flesh-and-blood truths whose presence I cannot deny" (Camus 1955/1983, p. 102).

References

Annas, George J., and Michael A. Grodin (eds.). 1992. *The Nazi doctors and the Nuremberg Code*. New York/Oxford: Oxford University Press.

Bayertz, Kurt. 1994. *GenEthics: Technological intervention in human reproduction as a philosophical problem*. Cambridge: Cambridge University Press.

Beecher, Henry K. 1959. *Experimentation in man*. Springfield: Charles C. Thomas.

Beecher, Henry K. 1966. Ethics and clinical research. *New England Journal of Medicine* 74: 1354–1360.

Blish, James. 1957. *The seedling stars and galactic cluster*. Hicksville: Gnome Press, Inc.

Burnett, Sir Macfarlane. 1978. *Endurance of life: The implications of genetics for human life*. London: Cambridge University Press.

Camus, Albert. 1955/1983. *The myth of sisyphus, and other essays*. New York: Vintage International/Random House, Inc.

Curran, William. 1982. Subject consent requirements in clinical research: An international perspective for industrial and developing countries. In *Human experimentation and medical ethics*, ed. Zbigniew Bankowski and Norman Howard-Jones. Geneva: Council for International Organizations of Medical Science.

Darwin, Charles. 1874/1974. *The descent of man and selection in relation to sex*. Chicago: Rand, McNally & Co.

Derrida, Jacques. 1978. Structure, sign and play in the discourse of the human sciences. In *Writing and difference*, ed. Jacques Derrida. Chicago: University of Chicago Press.

Dunn, Katherine. 1983. *Geek love*. New York: Warner Books.

Eccles, Sir John. 1970. *Facing reality*. Heidelberg/Berlin: Heidelberg Science Library/Springer.

Eccles, Sir John. 1979. *The human mystery (The Gifford lectures, 1977–78)*. Berlin/ Heidelberg/New York: Springer.

Elshtain, Jean Bethke. 1998. To clone or not to clone. In *Clones andclones: Facts and fantasies about human cloning*, ed. Martha C. Nussbaum and Cass R. Sunstein. New York/London: Norton.

Engelhardt Jr., H.T. 1973. The philosophy of medicine: A new endeavor. *Texas Reports on Biology and Medicine* 31: 443–452.

Gilbert, Walter. 1992. A vision of the Grail. In *The code of code: Scientific and social issues in the human genome project*, ed. Daniel J. Kevles and Leroy Hood, 83–97. Cambridge, MA: Harvard University Press.

Howard-Jones, Norman. 1982. Human experimentation in historical and ethical perspectives. *Social Science and Medicine* 16: 1429–1448.

Husserl, Edmund. 1965. *Phenomenology and the Crisis of Philosophy*. Trans. and Intro. by Quentin Lauer. New York: Harper Torchbooks/Harper & Row, Publishers.

Ingelfinger, Franz J. 1972. Informed (But uneducated) consent. *New England Journal of Medicine* 287: 465–466.

Jonas, Hans. 1966. *The phenomenon of life*. Chicago: University of Chicago Press.

Jonas, Hans. 1974. *Philosophical essays: From ancient creed to technological man*. Englewood Cliffs: Prentice-Hall, Inc.

Jonas, Hans. 1984. *The imperative of responsibility: In search of an ethics for the technological age*. Chicago: University of Chicago Press.

Jones, James H. 1981. *Bad blood: The Tuskegee Syphilis experiment*. New York: The Free Press.

Kass, Leon. 1997. The wisdom of repugnance. *The New Republic*, June 2.

Katz, Jay. 1992. The consent principle of the Nuremberg Code: Its significance then and now. In *The Nazi doctors and the Nuremberg Code*, ed. George J. Annas and Michael A. Grodin. New York/Oxford: Oxford University Press.

Laing, R.T. 1967. *The politics of experience*. New York: Pantheon Books.

Latour, B. 1988. *The pasteurization of France*. Cambridge, MA: Harvard University Press.

Lévi-Straus, Claude. 1967. *Les structures élémentaires de la parenté*, 2nd ed. The Hague: Mouton.

Markey, Edward J. (D, MA). 1986. *American nuclear guinea pigs: Three decades of radiation experiments on U. S. citizens*. A sub-committee staff report for the subcommittee on energy and power of the committee on energy and commerce, U.S. House of Representatives.

Mawer, Simon. 1998. *Mendel's Dwarf*. New York/London: Penguin Book.

Pellegrino, Edmund D. (ed.). 1991. *Ethics, trust, and the professions: Philosophical and cultural aspects*. Washington, DC: Georgetown University Press.

Report and Recommendations of the National Bioethics Advisory Commission. 1997. *Cloning human beings*, 19–23. Rockville: U.S. Department of Commerce.

Rescher, Nicholas. 1982. Moral issues relating to the economics of new knowledge in the biomedical sciences. In *New medical knowledge in the biomedical sciences*, ed. W.B. Bondeson et al., 33–45. Dordrecht/Boston: D. Reidel Publishing Co.

Rheinberger, Hans-Jörg. 1995. Beyond nature and culture: A note on medicine in the age of molecular biology. In *Medicine as a cultural system*, special issue of *Science in context* (8:1), ed. Michael Heyd and Hans-Jörg Rheinberger.

Schutz, Alfred. 1966. The problem of transcendental intersubjectivity in Husserl. In *Collected papers*, vol. III, ed. Alfred Schutz. The Hague: Martinus Nijhoff.

Schutz, Alfred. 1967. Scheler's theory of intersubjectivity and the general thesis of the alter ego. In *Collected papers*, vol. I, ed. Alfred Schutz. The Hague: Martinus Nijhoff.

Schutz, Alfred, and Thomas Luckmann. I: 1973. *Structures of the life-world*. Evanston: Northwestern University Press.

Spiegelberg, Herbert. 1961. Accident of birth: A non-utilitarian motif in Mill's philosophy. *Journal of the History of Ideas* 22: 10–26.

Spiegelberg, Herbert. 1974. Ethics for fellows in the fate of existence. In *Mid-century American philosophy: Personal statements*, ed. P. Bertocci, 193–210. New York: Humanities Press.

Tempkin, Oswei. 1973. *Galenism: Rise and decline of a medical philosophy*. Ithaca/London: Cornell University Press.

Weissmann, Gerald A. 1982. The need to know: Utilitarian and esthetic values of biomedical science. In *New medical knowledge in the biomedical sciences*, ed. W.B. Bondeson et al., 106–110. Dordrecht/Boston: D. Reidel Publishing Co.

Wilmut, I., A.E. Schnieke, et al. 1997. Viable offspring derived from fetal and adult mammalian cells. *Nature* 385: 810–813.

Wilmut, I., A.E. Schnieke, et al. 1998. Viable offspring derived from fetal and adult mammalian cells. In *Clones and clones: Facts and fantasies about human cloning*, ed. Martha C. Nussbaum and Cass R. Sunstein, 21–28. New York/London: W. W. Norton & Co.

Winner, Langdon. 2004. Resistance is futile: The posthuman condition and its advocates. In *Is human nature obsolete? Genetics, bioengineering, and the future of the human condition*, ed. Harold W. Baillie and Timothy K. Casey, 385–411. Boston: MIT Press.

Winner, Langdon. 2005. Resistance is futile: The post-human condition and its advocates. In *Is human nature obsolete? Genetics, bioengineering, and the future of the human condition*, ed. Harold W. Baillie and Timothy K. Casey. Boston: MIT Press.

Zaner, R.M. 2001a. Thinking about medicine. In *Handbook of phenomenology and medicine*, ed. A. Kay Toombs, 127–144. Dordrecht/Boston/London: Kluwer Academic Publishers.

Zaner, R.M. 2001b. Brave new world of genetics. Berlin: Duncker & Humblot, *Jahrbuch für Recht und Ethik* (*Annual Review of Law and Ethics*). (Band 9), 297–322.

Zaner, R.M. 2003. Finessing nature. *Philosophy and Public Policy Quarterly* 23(3), 14–19. Included in Verna Gehring (2003). (Ed.). *Genetic prospects: Essays on biotechnology, ethics and public policy*, 63–74. New York: Rowman, Littlefield Pub. Inc.

Zaner, R.M. 2005. Visions and re-visions: Life and the accident of birth. In *Is human nature obsolete? Genetics, bioengineering, and the future of the human condition*, ed. Harold W. Baillie and Timothy K. Casey, 177–207. Boston: MIT Press.

Zaner, R.M. 2013–2014. The indomitable Rachel Bittman, a story, in *Langdon Review of the Arts in Texas* 10, 68–79, preceded by "Why I write".

Printed in the United States
By Bookmasters